材料科学与工程专业英语配套教材

英汉电化学与表面处理专业词汇

王玥 郭晓斐 袁兴栋 编

ENGLISH-CHINESE DICTIONARY OF
ELECTROCHEMISTRY AND
SURFACE TREATMENT

化学工业出版社

·北京·

本书收集了电化学、电镀、涂装、物理涂镀、磷化、发蓝、氧化、化学热处理、涂镀三废处理、表面层理化测试技术及无损检测等专业词汇一万五千余条，书后备有附录（元素名称英汉对照表，常用电化学与表面处理缩写词，常见的构词形式）。

本书可作为高等学校材料专业、腐蚀防护专业以及电化学专业的专业英语配套教材，也可供相关领域科研人员、教学工作者、工程技术人员及职业学院学生查阅参考。

图书在版编目（CIP）数据

英汉电化学与表面处理专业词汇/王玥，郭晓斐，袁兴栋编 . —北京：化学工业出版社，2014.4
ISBN 978-7-122-19787-0

Ⅰ.①英… Ⅱ.①王… ②郭… ③袁… Ⅲ.①电化学-词汇-英、汉②金属表面处理-词汇-英、汉 Ⅳ.①O646-61 ②TG17-61

中国版本图书馆 CIP 数据核字（2014）第 027816 号

责任编辑：杨　菁　　　　　　　　　文字编辑：林　丹
责任校对：宋　夏　　　　　　　　　装帧设计：韩　飞

出版发行：化学工业出版社（北京市东城区青年湖南街 13 号　邮政编码 100011）
印　　装：三河市延风印装厂
787mm×1092mm　1/16　印张 14¾　字数 372 千字　2014 年 7 月北京第 1 版第 1 次印刷

购书咨询：010-64518888（传真：010-64519686）　售后服务：010-64518899
网　　址：http://www.cip.com.cn
凡购买本书，如有缺损质量问题，本社销售中心负责调换。

定　　价：49.00 元　　　　　　　　　　　　　　　　　版权所有　违者必究

前 言

表面技术是一门广博精深，极具实用价值的基础技术。其不仅是产品内在质量得以保证的关键，也是产品外观及性能得以保证的重要手段。表面技术涵盖了电镀、涂装、化学镀、磷化、氧化、化学热处理、物理涂镀等多种技术，广泛应用于冶金、机械、五金、电子、建筑、航天航空、兵器、能源、轻工及仪表等多个领域。表面处理技术的快速发展及广泛应用被认为是制造领域中的重要进展。随着科技的不断进步，表面技术新工艺、新方法、分析及检测手段不断涌现，外文文献的阅读量不断增大，国内外学术交流也更为频繁，为此，编写这本电化学与表面处理专业词汇，为读者提供一本阅读和翻译英文电化学、表面处理技术的工具书，提高英文阅读的效率及质量。

本书共收集15200余词目，内容以电化学、电镀、涂装、物理涂镀、化学热处理、磷化、轻金属及合金氧化及钢铁发蓝等表面处理技术词汇为主。同时，考虑阅读翻译专业技术读物的实际需要，也收入了材料表面成分分析、涂镀液三废处理、腐蚀与防护、无损检测及现代先进的理化检测手段方面的词汇。

在选定词汇进行编译时，编者查阅了大量的相关资料，特别注重译文的准确、精炼与完整，尽可能地结合相关标准、规范和教材，力求做到专业、简明、实用和展现最新技术。

参与本书编写的人员有山东建筑大学材料学院冯立明、孙华、项东、蔡元兴、孙齐磊、江荣岩、杜捷，齐鲁工业大学化学与制药工程学院杨鹏飞，济南晶恒电子（集团）有限责任公司齐建国、张华平，烟台三环锁业集团有限公司张庆来，香港中文大学机械学院樊帆，山东建筑大学材料学院张萌、代金山、孔德钰、周勇、于占举、贾腾、杨瑞康、冯加明、陶雪、王东芳、林伟丽、李王厚、李玉亮、张丹、郑文雪参与了文字的校验工作。

在本书的编写过程中得到刘科高博士的指导和帮助，在此表示感谢。

由于编者水平有限，在使用本《词汇》时如发现有疏漏或不当之处，敬请批评指正，衷心希望广大读者提出宝贵的意见。

编者

2014年1月

使用说明

1. 全部词条一律按照字母的顺序编排。
2. 不同词性的释义，用";"隔开。
3. 缩写字的"="后是所缩代的词的全文，例如词目 meq＝milligramequivalent 毫克当量。
4. 符号

① "词目-"代表其前的词目为构词成分，例如 oxa- 氧杂，其中 oxa 为构词成分。

② "［］"内的字可以各自替换前面那个（或几个）字，例如 demulitiplication 倍［缩，递］减＝①倍减 ②缩减 ③递减。

③ "（）"内的汉字是辅助释义，例如 desilicification 脱硅（作用，过程）；

"（）"内的小写英文字母表示可以省略的部分，例如 micell（a）胶束，胶态离子，微胞；

"（）"内的大写英文字母表示整个词条的缩写形式，例如 high density polyethylene（HDPE）高密度聚乙烯。

目 录

词汇正文 ··· 1
附录 ·· 219
　附录 1　化学元素名称英汉对照表（原子序数 100 号之前）············· 219
　附录 2　常见缩写词 ··· 221
　附录 3　常见的构词形式 ·· 226

A

A. C. 交流电
A. C. magnetic saturation 交流磁饱和
A. C. polarography 交流极谱法
A. C. voltage 交流电压
A. C. voltammetry 交流伏安法
A. C. superposed on D. C. plating 交直流叠加电镀
abate 减少，减退，降低，降低，消除
abaxial 轴外的，离开轴心的
abbe number 色散系数
abbertite 黑沥青
abbreviate 将……缩短，省略，约分
abbreviation 缩短，缩写，节略，简化
aberration 偏差，误差，色差，畸变，变形体
abeyance 潜态，暂时无效，中止，停顿
abietene 松香烯
abiochemistry 非生物化学，无机化学
abiotic degradation 非生物降解
ablate 消融，烧蚀，腐蚀，蒸发掉，风化，剥落
ablation 消融，烧蚀，风化，剥[脱]落，磨削，切除
ablution 洗净，清洗
abnormal 不规则的，不正常的，反常的
abnormal curve 不规则曲线
abnormal glow 不规则辉光放电
abnormal grain growth 反常晶粒生长
abnormal phenomena 反常现象
abnormal setting 异常凝结
abnormal steel 异常钢，反常钢
abnormal structure 反常组织[结构]
abnormalism 异常性，变态性
abnormality 反常性，不正常，不规则，变态
abradant 磨料，摩擦剂
abrade 磨损，耗损，磨蚀
abrader 磨损试验机，磨光机，砂轮机

abrasion 磨耗
abrasion cutting 磨切
abrasion loss 磨损量
abrasion resistance improvers 耐磨改性剂
abrasion resistance test 耐磨耗性试验
abrasion resistance 耐磨能力，耐磨耗性
abrasion test 磨耗试验
abrasion testing machine 磨损试验机
abrasive 磨损的，磨蚀的，磨料，研磨剂
abrasive action 磨蚀作用
abrasive belt 研磨带
abrasive brick 研磨砖
abrasive cloth 研磨布，砂布
abrasive disc 研磨盘，砂轮，磨轮
abrasive dresser 砂轮修整器
abrasive dust 磨屑
abrasive grain 磨料颗粒
abrasive hardness 磨蚀[耗]硬度，研磨硬度
abrasive industry 磨料工业
abrasive jet wear testing 喷砂磨损试验
abrasive machine 砂轮机
abrasive machining 磨削加工，强力磨削
abrasive material 磨料
abrasive media 研磨介质
abrasive paper 砂纸
abrasive paper sandpaper 打磨砂纸
abrasive resistance 抗磨力，抗磨性，耐磨能力
abrasive wear 磨耗，磨损，磨蚀
abrasive wheel 磨轮，砂轮
abrasiveness 磨损性，磨蚀性，磨耗
abrator 抛丸清理机，喷丸清理
abridge 删节，缩短，省略，简化
abridged drawing 略图
abridged general view 示意图
abrupt 突变[陡，不连续，生硬，突然]的

abrupt change 突变，陡变
abrupt curve 陡变曲线
abrupt junction 突变结，阶跃结
abruptly brittle rupture 突然脆性破坏
abscess （金属中）泡孔
abscissa 横坐标
absolute absorption 绝对吸收
absolute alcohol 无水酒精
absolute error 绝对误差
absolute potential 绝对电势
absolute rate theory 绝对速率理论
absolute temperature 绝对温度
absorbability 吸收性，吸收能力，吸收量
absorbance 吸光度
absorbate 被吸收物
absorbed dose 吸收剂量
absorbency 吸收性，吸光度，吸收能力，吸光率
absorbent 吸收剂，吸收质；能吸收的，有吸收能力的
absorbent carbon 活性炭
absorber 吸收器，吸收剂，（X射线）缓冲装置
absorbing 吸引人的，非常有趣的
absorbing agent 吸收剂
absorbing capacity 吸收能力
absorbite 活性炭
absorptance 吸收比
absorptiometer 吸收比色计
absorptiometric analysis 吸光分析
absorptiometry 吸收分光光度法
absorption 吸收；专注
absorption band 吸收带
absorption border 熔蚀边缘
absorption bottle 吸收瓶
absorption capacity 吸收能力
absorption cell 吸收池
absorption chamber 吸收室
absorption chromatography 吸收色谱法
absorption coefficient 吸收系数
absorption column 吸收柱

absorption constant 吸收常数
absorption edge 吸收限
absorption effect 吸收效应
absorption factor 吸收系数；吸收因素
absorption frequency 吸收频率
absorption impurity 吸收杂质
absorption index 吸收系数［指数，指标］
absorption jump 吸收跃迁
absorption law 吸收定律
absorption layer 吸收层，吸附层
absorption length 吸收长度
absorption lifetime 吸收寿期
absorption line 吸收线
absorption loss 吸收损失
absorption power 吸收能力
absorption probability 吸收概率
absorption rate 吸收率
absorption ratio 吸收系数
absorption reaction 吸收反应
absorption refrigeration 吸收制冷
absorption spectrometry 吸收光谱测定法
absorption spectrophotometry 吸收分光光度法
absorption spectroscopy 吸收光谱法
absorption spectrum 吸收光谱
absorption surface 吸收面
absorption test 吸收试验
absorption thickness 吸收厚度
absorption tower 吸收塔
absorptive 吸收的，吸水性的，有吸收力的
absorptive character 吸收性能
absorptive index 吸收指数，吸收系数，吸收率
absorptivity 吸收能力，吸收率，吸热率，吸湿性
abstention 戒除，弃权，节制
abstergent 洗净剂，去污粉；洗去……的，去垢的
abstersion 洗净，净化
abstersive 使……洁净的，去垢的

abstract 抽象，摘要；抽象［理论上］的；提取，分离，概括，摘录
abstract function 抽象函数
abstract heat 散热
abstract model 抽象模型
abstraction 夺取（反应），抽象，分离，提取，萃取
abstractly 抽象地，理论上
abstractness 抽象性
absurdity 不合理，谬论
abtragung 剥蚀作用
abundance 丰富，充裕，丰度
abundance anomaly 丰度变异
abundance measurement 丰度［分布量］测定
abundance zone 富集带
abundant 丰富的
abuse 滥用，违反操作规程
abutment joint 对接接头，平接缝，对接缝
abutment ring 连接环
abutting joint 对接接头
accelerant 促进剂
accelerated clarification 加速澄清
accelerated corrosion test 加速腐蚀试验
accelerated delamination test 快速层离试验
accelerated deterioration 加速老化
accelerated exposure test 加速大气腐蚀试验
accelerated flow 加速流
accelerated life test 加速寿命试验
accelerated light fastness test 耐晒度试验
accelerated load test 加速载荷试验
accelerated oxidation 加速氧化
accelerated test 加速试验
accelerated weathering test 加速风化试验，加速耐候性试验
accelerating action 加速作用
accelerating agent 促进剂，速凝剂
accelerating effect 加速作用

accelerator 加速剂，催速剂
accelerating field 加速场
acceleration 加速
acceleration detector 加速度检波器
acceleration effect 加速效应
acceptance level 验收水平，验收标准
acceptance limits 验收范围，容许极限
acceptance specification 验收规范
acceptance standard 验收标准
access 通路
accessory 附件，配件，附属设备
accidental error 随机误差
accumulate 累积，聚集，积累，存储，堆积
accumulated deformation 累积变形
accumulated error 累积误差
accumulation 累积，累加，积聚
accumulation layer 聚集层
accumulation of mud 淤泥
accumulation test 累积检测
accumulational 累积的，聚集的
accumulative 累积的，聚集的
accuracy 精确（性），准确（性），准确度
accuracy test 精度试验
accurate 准确的，精确的，精密的，已校准的
accurate die casting 精密压铸
accurate pointing 精确定向，点测
acerdol 高锰酸钙
acerose 针叶树的，针状的，针形的
acerous 针状的，针形的
acescent 容易变酸的，有酸味的，微酸的
acet 乙酰
acetal 乙缩醛
acetaldehyde 乙醛
acetamide 乙酰胺
acetate 醋酸盐
acetic 醋（酸）的
acetic acid 醋酸，乙酸
acetic acid glacial 冰醋酸

acetic acid salt spray test   醋酸盐雾试验
acetify   使醋化，使发酸
acetoacetate   乙酰醋酸酯，乙酰醋酸盐
acetone   丙酮
acetonic   丙酮的
acetonyl   丙酮基，乙酰甲基
acetoxy   醋酸基，乙酰氧基
acetyl   乙酰基
acetylacetone   乙酰丙酮
acetylate   乙酰化；乙酰化产物
acetylating agent   乙酰化剂
acetylene   乙炔
acetylide   炔化物
achromatic   消色的，无色的，非彩色的
achromaticity   消色差，无色，非彩色
achromatism   消色差，无色，非彩色
achromatization   消色差化，色差的消除
achromatize   使无色，消……色差，使成非彩色
achromatized   消色差的
achromic   消色的，无色的
achromous   消色的，无色的
aci-compound   酸式化合物
acicular   针状的
acicular powder   针状粉末
acicular crystal   针状晶体
acicular structure   针状组织，贝氏体
aciculate   针状的
acid   酸
acid anhydride   酸酐
acid-base   酸碱的
acid base equilibrium   酸碱平衡
acid base indicator   酸碱指示剂
acid base neutralization titration   酸碱中和滴定
acid base pair   酸碱对
acid base titration   酸碱滴定
acid bath   酸浴
acid brittleness   酸洗脆性
acid bronze   耐酸青铜
acid burette   酸式滴定管

acid catalysts   酸催化剂
acid cleaning   酸洗
acid copper plating   酸性镀铜
acid-deficient   弱酸的，缺酸的
acid degreasing   酸脱脂
acid dipping   弱侵蚀
acid embossing   酸刻
acid-etched   酸侵蚀的，酸蚀刻的
acid etching   酸蚀
acid-fast   耐酸的，抗酸性的
acid-free   无酸的
acid flux   酸性溶剂
acid group   酸根
acid hydrolysis   （加）酸（水）解（作用）
acid lining   酸性衬里
acid nitrile   腈
acid number   酸值
acid pickle   废酸液
acid pickling   酸洗
acid pickling inhibitor   酸洗阻蚀剂
acid polishing   酸磨光
acid proof alloy   耐酸合金
acid proof cast iron   耐酸铸铁
acid radical   酸根（基），酰
acid reaction   酸性反应
acid refractories   酸性耐火材料
acid resistance   耐酸性
acid resisting steel   耐酸钢
acid salt   酸性盐
acid scavengers   酸清除剂
acid slag   酸性渣
acid sludge   废酸
acid soluble   酸溶的，可溶于酸的
acid solution   酸溶液
acid solvent   酸性溶剂
acid treated   酸化的，酸处理过的
acid treatment   酸处理
acid value   酸值
acidamide   酰胺
acidate   酸（酰）化
acidation   酸化，酰（基取）代

acidic 酸（性）的，酸式
acidic cleaner 酸洗清洁剂
acidic component 酸性成分
acidic titrant 酸性滴定剂
acidic oxide 酸性氧化物
acidiferous 含酸的
acidifiable 可酸化的
acidification 酸化（作用），发酸
acidimeter 酸比重计
acidimetry 酸量滴定法
acidity 酸度
acidometer 酸度计
acidometry 酸度测定法
acidproof 耐酸性的，耐酸的
acidproof paint 耐酸漆
acidylable 酰化的
acidylate 酰化，使酰代
acierage 表面钢化
aciform 针状的
acinose 细粒状的
acoustic emission (AE) 声发射
acoustic emission count 声发射计数
acoustic emission transducer 声发射换能器
acoustic holography 声全息术
acoustic impedance 声阻抗
acoustic impedance matching 声阻抗匹配
acoustic impedance method 声阻法
acoustic wave 声波
acoustic-ultrasonic 声-超声
across 交叉，横过
acryl 丙烯（醛基）
acryl lacquer 热塑性丙烯酸涂料
acryl modified slkyd resin 丙烯酸改性的醇酸树脂
acrylaldehyde 丙烯醛
acrylamide 丙烯酰胺
acrylate 丙烯酸盐［酯］
acrylate resins 丙烯酸系树脂
acrylic 丙烯酸的，聚丙烯的
acrylic acid 丙烯酸
acrylic resin 丙烯酸树脂

acrylic resin paint 丙烯酸树脂涂料
acrylonitrile butadiene styrene resin (ABS) 丙烯腈-丁二烯-苯乙烯树脂
activated adsorption 活性吸附
activated atom 活化原子
activated carbon 活性炭
activated complex 活化络合物
activated complex theory 活化配［络］合物理论
activated molecule 活化分子
activated metals 活化金属
activated silica 活性硅
activated sludge process 活性污泥法
activated sludge 活性污泥
activating group 活化基团
activation 活化，活性化处理
activation energy 激化能，活化能
activation polarization 电化学极化，活化极化
activator 活化剂，催化剂
active 活性的，活化的
active carbon 活性炭
active carbon adsorption 活性炭吸附
active center 活化中心
active deposit 活性沉积物
active hydrogen compounds 活泼氢化合物
active intermediate 活性中间体
active material utilization 活性物质利用率
active material 放射（性）材料，活性物质
active overpotential 活化过电位
active-passive metal 活性-钝态金属
active-passive transition 活性-钝态转变
active region 活性区
active zone 活性区，活化区
activity 活度
activity coefficient 活度系数
actual capacity 实际容量
acyl cation 酰（基）正离子
acyl chloride 酰氯
acyl cyanide 酰腈

acyl fluoride 酰氟
acyl group 酰基
acyl halide 酰卤
acyl iodide 酰碘
acyl peroxide 酰基过氧化物
acyl tosylate 酰基对甲苯磺酸酐
acylation 酰化
acyloxyation 酰氧基化
adaptability 适应性
adaptation 适应
addition 附加
addition agent 添加剂
addition polymer 加聚物
addition polymerization 加聚反应
addition reaction 加成反应
additional 附加的
additional stress 附加应力
additive wear 黏着磨损
additive 添加剂
adequate shielding 适当防护，适当屏蔽
adherence 黏合
adherence test 黏着试验
adhesion 黏附，结合力，附着力
adhesion force 附着力
adhesion promoter 附着力促进剂
adhesion strength 附着强度
adhesion test 附着性测试
adhesive 胶黏的；黏着剂
adhesive capacity 黏着能力
adhesive film 黏附膜
adhesive power 黏附力
adhesivity 黏附性
adiethyl ether 乙醚
adion 吸附离子
adipic acid 己二酸
adjustable 可调整的，可调节的
admittance 电纳，导纳，容差
admittance test 导纳试验
admixture 掺和物
adsorbability 吸附性
adsorbate 吸附质，被吸附物

adsorbed film 吸附膜
adsorbed layer 吸附层
adsorbent 吸附剂
adsorber 吸附器
adsorption 吸附
adsorption analysis 吸附分析
adsorption capacity 吸附能力
adsorption compound 吸附化合物
adsorption current 吸附电流
adsorption energy 吸附能
adsorption equilibrium 吸附平衡
adsorption exponent 吸附指数
adsorption heat 吸附热
adsorption indicator 吸附指示剂
adsorption isotherm 吸附等温线
adsorption theory of multi-molecular layers 多分子层吸附理论
adsorption water 吸附水
adsorption wave 吸附波
adsorptive stripping voltammetry 吸附溶出伏安法
aerated water 充气水
aeration 曝气，充气，吹风
aeration cell 充气电池
aerator 曝气设备，通风装置
aerobe 需氧微生物
aerobic bacteria 嗜氧细菌
aerobic digestion 需氧消化
aerobic sludge digestion 需氧污泥消化
aerobic 需氧的，好氧的（细菌）
aerosil 硅胶
aerospace corrosion 太空腐蚀
aerospace material 太空材料
aerugo 铜绿
affinity 亲和力，吸引力
after precipitation 后沉淀
after tack 返黏
after treatment 后处理
agar 琼脂
age hardening 时效硬化，时效强化
ageing 老化处理，时效

ageing treatment 时效处理
agent 试剂
agent of oxidation 氧化剂
agglomerated structure 聚结组织
agglomerating 黏聚，凝聚，聚结
agglomeration 结块，成团
agglutination 凝集
agglutination reaction 凝集反应
aggregation 聚集
aggressivity 侵蚀性
aging action 老化作用
aging crack 时效裂纹（痕）
aging property tester 老化性能测定仪
aging range 时效温度范围
aging test 时效试验
agitation 搅动，摇动
agitator 搅拌机，搅拌器
air blast 鼓风
air blast quenching 强风淬火
air blow 吹风
air breakdown 空气绝缘破坏
air bubble 气泡
air cap 空气帽
air channel 开路
air cleaner 空气净化器
air compressor 空气压缩机
air condenser 空气冷凝器
air cooling 气冷
air dryer 空气干燥器
air drying cold curing 自然干燥，常温干燥
air drying 空气干燥
air duster 气动除尘器
air hardening 气体硬化
air header 集气管
air hole 气孔，风眼
air humidity 空气湿度
air knife 气刀
air machine 通风机
air meter 气量计
air nozzle 空气喷嘴

air oxidation 空气氧化
air patenting 空气韧化
air permeability 透气性
air pollution 大气污染
air port 通气口
air pressure 气压
air pump 气泵
air purifier 空气净化器
air sander 气动打磨机
air seal 风幕，气封
air seasoning 天然干燥法
air set 空气中凝固，常温自硬，自然硬化
air shrinkage 风干收缩
air spraying 空气喷涂
air supply 气源
air supply house 空调装置
air tightness test 气密试验
air transformer 空气压力调整器
airless spraying equipment 无气喷涂装置
airless blast cleaner 离心喷砂机
airless blast cleaning 离心喷光法
airless plasting cleaning 离心喷光
airless spray gun 无气喷枪
airless spraying 无气喷涂，压力喷涂
alarm condition 报警状态
alarm 警报器
alarm level 报警电平
albuminoid nitrogen 蛋白性氮
alcohol ether carboxylate（AEC） 醇醚羧酸盐
alcohol 醇
alcoholic compound 醇化合物
alcoholic solution 醇溶液
alcohols solvent 醇系溶剂
alcoxyl 烷氧
aldehyde 醛
aldehyde acid 醛酸
aldehyde alcohol 醛醇
aldol 羟醛
aldol condensation 醛醇缩合
aldose 醛糖

algaecide 杀藻剂
algicide 杀藻剂
algistat 抑藻剂
alicyclic 脂环族的
alicyclic compound 脂环化合物
alignment accuracy 位置对准精确度
alignment mark 对准标记
aliphatic 脂肪族的
alignment 定向，对准，定位调整，校直
aliphatic acid 脂肪族酸
aliphatic alcohol 脂肪族醇
aliphatic amine 脂肪族胺
aliphatic base 脂肪族碱
aliphatic compound 脂肪族化合物
aliphatic ether 脂肪族醚
aliphatic hydrocarbon 脂肪族烃，链烃
aliphatic series 脂肪族系
aliphatic unsaturated carboxylic acid 脂肪族不饱羧酸
aliquation 偏析，熔析，层化，起层
alizarin 茜素，1,2-二羟基蒽醌
alizarin blue 茜素蓝
alizarin brown 茜素棕
alizarin dye 茜素染料
alizarin lake 茜素色淀
alizarin yellow 茜黄
alizarine 茜素
alkali 碱
alkali burette 碱式滴定管
alkali cellulose 碱纤维素
alkali cleaning 碱洗
alkali copper 碱铜
alkali dipping 脱脂
alkali embrittlement 碱脆性
alkali liquor 碱液
alkali metals 碱金属
alkali resistance 耐碱性
alkali salt 碱金属盐
alkalimetric 碱量滴定的
alkalimetry 碱量滴定法
alkaline battery 碱性电池

alkaline cell 碱性电池
alkaline cleaner 碱性清洗剂，碱液脱脂剂
alkaline cleaning 碱洗
alkaline derusting 碱法除锈
alkaline degreasing 化学除油，碱液脱脂，碱蚀
alkaline earth metal 碱土金属
alkaline earths 碱土族
alkaline reaction 碱性反应
alkaline solution 碱性溶液
alkalinity 碱度[性]
alkaliproof 耐碱性
alkali-proof paint 耐碱漆
alkalization 碱化
alkaloid 生物碱，植物碱基
alkamine 氨基醇类
alkane 链烷，烷烃
alkanolamine 烷烃醇胺
alkansulfonic acid 链烷磺酸
alkene 链烯，烯烃
alkine 链炔，炔属
alkyd resin 醇酸树脂
alkyd resin paint 醇酸树脂涂料
alkyl 烷基，烃基；烷基的，烃基的
alkyl amido phosphate ester 烷基酰胺磷酸酯
alkyl cellulose 烷基纤维素
alkyl group 烷基
alkyl halide 烷基卤，卤代烷
alkyl phosphate (AP) 烷基磷酸酯
alkyl polyglycoside (APG) 烷基糖苷
alkyl sulfate 烷基硫酸盐
alkyl sulfonic acid 烷基磺酸
alkylate 烷基化产物
alkylating agent 烷化剂
alkylation 烷基化
alkylbenzene 烷基苯
alkylbenzene sulfonate 烷基苯磺酸盐
alkylogen 烷基卤，卤代烷
alkylolamide 烷基醇酰胺
alkylpheol ethoxylates 烷基酚聚氧乙烯醚

alkyne 炔烃
alkynol 炔醇
all pattying 统刮腻子
all purpose machine 万能工具机
allene 丙二烯
alligation （金属的）熔合，合金，混合法，和均性
alligator 颚口工具，颚式破碎机，鳄鱼皮；龟裂
alligatoring 鳄嘴裂口，（轧制表面）裂痕，龟裂
alligateying crocodiling 鳄鱼皮状裂纹
allomeric 异质同晶的
allomorphous 同质异晶的
allotrope 同素异形体
alloy 合金
alloy analysis 合金分析
alloy carbide 合金碳化物
alloy cast iron 合金铸铁
alloy element 合金元素
alloy plating 合金电镀
alloy steel casting 合金钢铸件
alloy strip 合金钢带
alloy steel 合金钢
alloy tool steel 合金工具钢
allowable minimum voltage 允许最小电压
allowable variation 允许偏差，容许变化
allumen 锌铝合金
allyl 烯丙基
allyl alcohol 烯丙醇
allyl cation 烯丙基正离子
allyl complex 烯丙基络合物
allyl compound 烯丙基化合物
allylene 丙炔
almasilium 硅镁铝合金（镁1%，硅2%，其余为铝）
alminal 铝硅合金
alpha brass α-黄铜
alpha bronze α-青铜
alpha cellulose α-纤维素
alpha coefficient 酸效应系数

alpha iron α-铁
alpha rays α-射线
alpha ray spectrometer α-射线能谱仪
altered layer 改性层
alternate immersion test 反复浸渍实验，交替浸渍实验
alternate wet and dry test 交替干湿试验
alternating bend test 反复弯曲试验
alternating copolymer 交替共聚物
alternating current anodizing 交流阳极氧化
alternating stress 交替应力，交变应力
alternation 交替
altitude 高度
alum 明矾
alumina 氧化铝，矾土
alumina fiber 氧化铝纤维
aluminium 铝
aluminium alloy 铝合金
aluminium alloy casting 铝合金铸件
aluminium ammonium sulfate 硫酸铝铵
aluminium casting 铸铝合金
aluminium coating 热镀铝法
aluminium flake paste 铝粉膏
aluminium flake powder 铝粉
aluminium foil 铝箔
aluminium hydroxide 氢氧化铝
aluminium impregnation 铝化，渗铝
aluminium mottling 银粉不均
aluminium nitrate 硝酸铝
aluminium oxide 氧化铝
aluminium pigment paste 铝粉膏
aluminium plate 铝板
aluminium potassium sulfate 硫酸铝钾，明矾
aluminium powder 铝粉
aluminium silicate 硅酸铝
aluminium sulfate 硫酸铝
aluminothermy 铝热法，铝冶术
aluminizing 渗铝
amalgam 汞齐
amalgam cell 汞齐电池

amalgam electrode 汞齐电极
amidation 酰胺化
amide 酰胺，氨基化合物
ambient temperature 环境温度，背景温度，周围介质温度，周围温度
amendment 修正
amide formation 酰胺的生成
amination 胺化，氨基化
amine 胺
amine content 胺浓度
amino acid 氨基酸
amino resin 氨基树脂
aminopyridine 氨基吡啶
aminoquinoline 氨基喹啉
aminosalicylic acid 氨基水杨酸
aminosulfonic acid 氨基磺酸
ammonia 氨
ammonia gas 氨气
ammonia water 氨水
ammonification 氨化（作用），加氨（作用）
ammonium acetate 乙酸铵
ammonium bifluoride 氟化氢铵
ammonium carbonate 碳酸铵
ammonium chloride 氯化铵，卤砂
ammonium dichromate 重铬酸铵
ammonium fluoride 氟化铵
ammonium formate 甲酸铵
ammonium hydrogen carbonate 碳酸氢铵
ammonium hydroxide 氢氧化铵
ammonium iron sulfate 硫酸铁铵
ammonium metavanadate 偏钒酸铵
ammonium molybdate 钼酸铵
ammonium nitrate 硝酸铵
ammonium oxalate 草酸铵
ammonium perchlorate 高氯酸铵
ammonium persulfate 过硫酸铵
ammonium phosphate 磷酸铵
ammonium phosphite 亚磷酸铵
ammonium phosphomolybdate 磷钼酸铵
ammonium rhodanide 硫氰酸铵
ammonium salt 铵盐

ammonium sulfate 硫酸铵
ammonium sulfite 亚硫酸铵
ammonium 铵
amorphism 非晶
amorphous 无定形的，无定形，非晶（型）的
amorphous carbon 无定形碳
amorphous graphite 无定形石墨
amorphous material 无定形材料，非晶态材料
amorphous metal 无定形金属
amorphous polymer 非晶态聚合物
amorphous state 非晶质状态
amount of oxygen precipitation 析出氧气量
amount of substance 物质的量
amount 数量
ampere 安培
amperometric titration 安培滴定法，电流滴定法
amphiphile 两亲物
amphiphilic 两性分子的
ampholyte 两性电解质
ampholytic active agent 两性表面活性剂
amphoteric compound 两性化合物
amphoteric element 两性元素
amphoteric oxide 两性氧化物
amphoteric ion 两性离子
amphoteric solvent 两性溶剂
amphoteric surfactant 两性表面活性剂
amplifier panel 放大器面板
amplitude 振幅、幅度
angle beam method 斜射法、角波束法
amyl 戊基
anaerobic 厌氧的（细菌）
anaerobic bacteria 厌氧菌
anaerobic corrosion 厌氧菌腐蚀
analog devices 模拟器件
analog-to-digital converter 模-数转换器
analogue 相似物，类似情况
analysis 分析

analysis area   分析区域
analysis with ion selective electrodes   离子选择电极分析法
analyte   被分析物，分解物
analytical balance   分析天平
analytical chemistry   分析化学
analytical method   分析法
analytical reaction   分析反应
analytically pure   分析纯
analyzer   分析器，分析仪，析光镜
anelastic creep   滞弹性潜变
anelastic deformation   滞弹性变形，塑性变形
anelastic strain   滞弹性应变
anelasticity   滞弹性
anhydrous   无水的
angle of emission   发射角
angle of incidence   入射角
angle of scattering   散射角
angstrom unit   埃（长度单位）
anhydr-   【构词成分】脱［去，无］水
anhydride   酸酐，脱水物
anhydrous   无水的
anhydrous acid   无水酸
anhydrous alcohol   无水酒精
anhydrous ammonia   无水氨
anhydrous salt   无水盐
aniline   苯胺
aniline black   苯胺黑
aniline blue   苯胺蓝
aniline dye   苯胺染料
animal glue   动物胶
animal oil   动物油
anion   阴离子，负离子
anion active agent   阴离子表面活性剂
anion coordination polyhedron   负离子配位多面体
anion exchange   阴离子交换
anion exchange membrane   阴离子交换膜
anion surfactant   阴离子表面活性剂
anionic polymerization   阴离子聚合

anionoid reagent   类阴离子试剂
anisaldehyde   茴香醛
anisotropic   各向异性的
anisotropic etching   各向异性蚀刻，非等向性蚀刻
anisotropic substance   异向性物质
anisotropy   各向异性，非均质性
annealing   退火
annealing box   退火箱
annealing embrittlement   退火脆性
annealing furnace   退火炉
annealing point   退火点
annealing temperature   退火温度
annealing texture   退火织构
annealing twin   退火双晶［孪晶］
annealing uniformity   退火处理之均质性
anode   阳极
anode bag   阳极袋
anode-cathode distance   极间距
anode coating   阳极镀层
anode corrosion efficiency   阳极腐蚀效率
anode coupling   阳极耦合
anode dissolution   阳极分解
anode earth method   阳极接地法
anode effect   阳极效应
anode efficiency   阳极效率
anode electrodeposition   阳极电沉积
anode electrocleaning   阳极电解除油
anode pickling   阳极浸渍（法）
anode polarization   阳极极化
anode process   阳极过程
anode slime   阳极泥
anode sludge   阳极泥
anodic current   阳极电流，阳极波
anodic current density   阳极电流密度
anodic inhibitor   阳极缓蚀剂
anodic oxidation   阳极氧化
anodic oxide coating   阳极氧化膜
anodic oxide film   阳极氧化膜
anodic polarization   阳极极化
anodic protection   阳极保护

anodic reaction 阳极反应
anodic spark deposition (ASD) 阳极火花沉积
anodic stripping voltammetry (ASV) 阳极溶出伏安法
anodic wave 阳极波
anodising 阳极处理，阳极化，阳极氧化
anodization 阳极化
anodization of sand blast 喷砂阳极处理
anodizing 阳极处理，电化学氧化
anolyte 阳极（电解）液
anomalous 不规则的，不协调的，不恰当的
antagonistic effect 对抗效应
antagonium 拮抗作用
anti blocking agent 防粘连剂
anti brass 青古铜
anti brass with extinction 消光青古铜
anti foaming agent 消泡剂
anti reflection coating 防反射涂膜
anti-corrosion formulation 防腐配方
anti-corrosion zinc alloy plating 耐蚀锌合金电镀
anti-corrosive paint 防腐漆
anti-crawling agent 防蠕变剂
anti-cratering agent 防缩孔剂
anti-float agent 防发花剂
anti-flooding agent 防浮色剂
anti-foamer 消泡剂
anti-fouling agent 防污剂
anti-freezing agent 防冻剂
anti-rust agents 防锈剂
anti-settling agent 防沉剂
anti-silking agent 防走丝剂
anti-skid agent 防滑剂
anti-skinning agent 防结皮剂
anti-static agent 抗静电剂
anti-tarnish 防褪色，防变色
anti-thixotropy 反触变性
antichlor 脱氯剂

anticorrosion 抗腐蚀
anticorrosion treatment 防腐蚀处理
anticorrosion wax 防锈蜡
anticorrosive agent 防腐蚀剂
antiferromagnetism 反铁磁现象，反铁磁性
antifoaming agent 消泡剂
antimony 锑
antioxidant 抗氧化剂
antique 古的
antique brass 青古铜
antique copper 红古铜
antirust paint 防锈漆
antirust wax 防锈蜡
antisag agent 流挂防止剂
antiseptics 防腐剂
antisetlle agent 沉淀防止剂
antistatic agent 静电防止剂
apatite 磷灰石
apparatus 仪器，设备，装置
apparent activation energy 表观活化能
apparent density 表观密度
apparent density of film 氧化膜表面密度
apparent equilibrium 表观平衡
apparent specific gravity 表观密度
apparent viscosity 表观黏度
apparent volume 表观体积
appearance 外观
appearance and surface finish 外观及表面精饰
appearance inspection 外观检查
appearance of paint film 涂膜的外观
appendix 附件
application by (dark) infrared drying 红外干燥
application by roller 辊涂
application drawing 操作图，应用图
application solid 施工固体分
application trial 试涂装
applied chemistry 应用化学
applied potential 外加电位

applied thermodynamics 应用热力学
appreciable 可预见的，可估计的
approach curve 渐近曲线
approved by 经……核准
approach curve 渐近曲线
approximate 近似的，大概的
approximate calculation 近似计算
approximate value 近似值
approximation 近似法，概算
appurtenances 附属设备
aprotic （对）质子惰性的，无施受的
aprotic solvent 非质子溶剂，无质子溶剂
aqua ion 水合离子
aqua regia 王水
aquagel 水凝胶
aqueous 含水的，水的
aqueous corrosion 水溶液腐蚀
aqueous solution 水溶液
aqueous vapor 水蒸气
arabic gum 阿拉伯胶
arbitration analysis 仲裁分析
arborescent powder 树枝状粉末
areal resistance 面电阻
areometer 比重计
areometry 比重测定法
argentometric method 银量法
argentometry 银量滴定，银量法
argon 氩
argon welding 氩焊
aroma 香味
aromatic 芳族的
aromatic acid 芳族酸
aromatic aldehyde 芳香醛
aromatic amine 芳香胺
aromatic compound 芳族化合物
aromatic hydrocarbon 芳香烃
aromatic ketone 芳香酮
aromatic nucleus 芳香环
aromatic polyamide 芳香聚酰胺
aromatic series 芳香系
aromatization reaction 芳香化反应
aroylation 芳酰基化
Arrhenius activation energy 阿累尼乌斯活化能
Arrhenius equation 阿累尼乌斯方程
Arrhenius ionization theory 阿累尼乌斯电离理论
arsenic 砷，三氧化二砷，砒霜；砷的，含砷（主要指五价砷）的
arsenic acid 砷酸
arsenic trioxide 三氧化二砷
artificial 人工的
artificial abrasive 人造磨料
artificial aging 人工时效
artificial corundum 人造金刚砂
artificial defect 人工缺陷
artificial diamond 人造金刚石
artificial fiber 人造纤维
artificial seawater 人造海水
artificial weathering 人工耐候性试验
aryl compound 芳基化合物
aryl disulfides 芳基二硫化物
aryl halide 芳基卤
arylamine 芳基胺
arylation 芳基化
arylide 芳基化物
aryloxy compound 芳氧基化合物
arylsulphonate 芳基磺酸盐
asbestine 石棉的，不燃性的
ascorbic acid 抗坏血酸
asepsis 防腐（法），无菌（法，操作）
ash content 灰分
ash marking 灰印
aspect ratio 纵横比
asphalt（bitumen） 沥青
assembly 装配
assimilative 同化的
assimilative organic carbon 可生化有机碳
assistant 助剂
associate 缔合
associated 缔合的
association 缔合

association chemical adsorption 缔合化学吸附
associative thickeners 缔合性增稠剂
assortment 分类
assembly 组装
asymmetric 不对称的
asymmetric A. C. plating 不对称交流电镀
asymmetric membrane 非对称膜
asymmetric oxidation 不对称氧化
asymmetric structure 不对称结构
asymmetric synthesis 不对称合成
asymmetrical 不对称的
asymmetrical stretching vibration 不对称伸缩振动
asymmetry 不对称
asymmetry parameter 非对称参数
atactic 不规则的，无规立构的
athermal transformation 非热转变
atmosphere 大气压
atmospheric air 大气空气
atmospheric corrosion 大气腐蚀
atmospheric corrosion test 大气腐蚀试验
atmospheric exposure test 大气暴露试验
atmospheric pollution 大气污染
atom 原子
atom percent 原子百分比
atomic absorption spectrometry (AAS) 原子吸收分光光度法，原子吸收光谱
atomic charge 原子电荷
atomic core 原子核
atomic emission spectrometer (AES) 原子发射光谱仪
atomic energy 原子能
atomic fluorescence spectrometry 原子荧光光谱法
atomic fluorometry 原子荧光分析法
atomic force microscope (AFM) 原子力显微镜
atomic group 原子团
atomic hydrogen 原子氢
atomic mass unit 原子质量单位
atomic mixing 原子混合
atomic orbital 原子轨道
atomic packing factor 原子填充因子
atomic radius 原子半径
atomic spectrum 原子光谱
atomic structure 原子结构
atomic symbol 原子符号
atomic unit 原子单位
atomic vibration 原子振动
atomic volume 原子体积
atomic weight 原子量
atomicity 原子数
atomized metal powder 雾化金属粉
atomizing 雾化
attenuation coefficient 衰减系数
attenuation constant 衰减系数
attenuation length 衰减长度
attrition 磨损
Auger electron spectroscopy (AES) 俄歇电子能谱
Auger electron spectrum 俄歇电子谱
Auger neutralization 俄歇中和
Auger parameter 俄歇参数
Auger transition 俄歇跃迁
augular offset 角度偏差
ausforming 形变淬火，形变热处理
austempering 奥氏体等温淬火
austenite 奥氏体
austenitic stainless steel 奥氏体不锈钢
austenitizing 奥氏体化
auto absorber 自动吸水器
auto roll filter 自动滚筒式过滤器
autocatalytic plating 化学镀，自催化镀
autoclave test 热压膨胀试验
automatic air duster 空气自动除尘器，自动吹风装置
automatic aluminum membrane shaper 自动冲膜机
automatic analysis 自动分析
automatic balance 自动天平
automatic buret 自动滴定管

automatic chemical diluting and mixing equipment 自动化学稀释混合设备
automatic control 自动控制
automatic damper/exhaust for coater 自动风门/排气涂料器
automatic dust off machine 自动除尘装置
automatic electrostatic spray machine 自动静电涂装装置
automatic elongation control 自动延伸控制
automatic gauge control 自动仪表控制
automatic machine 自动机械，自动装置
automatic sanding machine 自动湿打磨装置
automatic solution machine 自动溶液管理装置
automatic spray 自动涂装
automatic spray machine 自动喷涂机
automatic spraying system 自动喷淋系统
automatic temperature recorder 温度自动记录器
automatic testing 自动检测
automatic thermoregulator 自动温度控制器
automatic titration 自动滴定
automatic titrator 自动滴定仪
automatic washing machine 自动水洗装置
automation 自动化
autoradiography 自动射线照相术
autoxidation 自氧化
auxiliary anode 辅助阳极
auxiliary cathode 辅助阴极
auxiliary electrode 辅助电极
auxochrome 助色团
availability 利用度，利用率
available chlorine 有效氯
average 平均
average deviation 平均偏差
average voltage 平均电压
Avogadro's law 阿伏伽德罗定律
axiality 同轴度，轴对称性
azo violet 偶氮紫

B

back-feed 反馈
back flow 逆流，回流
back titration 回滴定，反向滴定
back wash 反洗，回洗
background limits 背景极限，电势窗
background signal 本底信号，背景信号
backing material 底材，基底材料
backlining 背衬，衬板
backscattering 反向散射，背反射
backscattering factor 反向［背］散射因子
backscattering spectrum 反向［背］散射谱
backswing 回复，回摆，反冲
backswing voltage 反向电压
bacteria corrosion 细菌腐蚀，微生物腐蚀
bactericidal 杀菌的
bactericide 杀菌剂
bacteriostat 抑菌剂
bacteriostatic 阻止细菌繁殖法的，抑制细菌的
bacteriostatic action 抑菌作用
baffle 挡板；用挡板控制
baffled 带有挡板的
baffler 折流板，调节墙
baffling 起阻碍作用的，阻碍……的
bag filter 袋滤器，袋式过滤器
bainite 贝氏体
bake 烤；被烤干，烘焙
bake schedule 干燥时间
bakeable 可烘烤的
baked 烤的，烘焙的；烘培
bakelite 酚醛塑料，电木
baking 烘干
baking enamel 焗漆
baking finish 烘漆，烤漆
baking oven 烘干室，烘箱
baking soda 小苏打，碳酸氢钠
baking stoving 烘干
baking varnish 烤漆

balance 平衡
balance pan 天平盘
balance rider 游码
balancing tank 均衡池，平衡槽
ball anode 球形阳极
ball bearing slide 脚轮滑道，滑轨
ball blast 抛丸，喷丸
ball blasting 喷丸清理
ball burnishing 钢珠滚光，钢珠抛光
ball cratering 球状弧坑
ball float valve 浮球阀；浮子阀
ball gauge 球形量规
ball grinder 球磨床，球磨机，磨球机
ball hardness 球印硬度，钢球硬度，布氏球测硬度，布氏球印硬度
ball hardness machine 钢球硬度试验机
ball hardness number 布氏硬度值
ball hardness testing machine 布氏硬度试验机
ball indentation 钢球压痕
ball indentation testing apparatus 球印硬度试验机
ball method of testing （测定硬度的）钢球压印试验
ball mill 球磨机
ball mill pulverizer 球磨粉碎机
ball mill refiner 球磨精研机
ball milling 球磨
ball non-return valve 单向球阀
ball nozzle 球形喷嘴
ball pendulum test 球摆试验
ball plug 球阀
ball relief valve 减压球阀，安全球阀，释放球阀，球形安全阀
ball spray treatment 喷丸处理
ball tester 球硬度试验机
ball valve 球阀，球形单向阀，球形止逆阀

ball-and-stick model 球棒模型
ball-cock 浮球阀
ball-cock assembly 浮球阀装置
ball-float liquid level meter 球形浮子液面计
ball-indentation hardness 球压硬度
ball-shooting 喷丸清理
band 带，条，频带，波段，条纹
band bending 能带弯曲
band drier 带式干燥机
band electrode 带状电极
band filter 带式过滤机，带通滤波器
band gap 带隙
band meter 波长计
band pickling machine 带材酸洗机
band separation 能带间距
band splitting 谱带分裂
band sprayer 带状喷雾机
band switching 波段变换
band-type papering machine 带式砂纸打磨机
banded structure 条纹状组织
bar 棒材，汇流条
bar chart 条形图，柱状图
bar section 型材，棒形截面
barbotage 鼓泡，起泡作用
bare 空的，赤裸的，无遮蔽的；露出
bare cable 裸电缆
bare metal 裸金属
baria 重晶石，氧化钡
baric 钡的，含钡的
barilla 苏打灰
baring 暴露，掘开
barite 重晶石（主要成分硫酸钡）
barium 钡
barium carbonate 碳酸钡
barium hydroxide 氢氧化钡
barium sulfate 硫酸钡
baro- 【构词成分】气压，压，重
barrel 桶，卷筒

barrel burnishing 滚光
barrel finishing 滚光
barrel plating 滚镀
barrel polishing 滚筒抛光
barrel tumbling 滚筒打光
barrelled 桶装的，装了桶的；把……装入桶内
barrelling 装桶，滚光加工，转桶清砂法
barren 贫瘠的，贫乏的
barren flux 净（不含金属的）熔剂
barricade 隔板，屏蔽；阻塞，遮住
barrier 障碍物，屏障，界线
barrier layer 阻挡层
barrier layer anodic oxide coating 阻挡层阳极氧化膜
barrier separation 膜分离
basal 基部的，基础的
base 碱
base bullion 粗金属锭
base coating 底基涂层
base electrolyte 基底电解质
base layer 底层
base metal 碱金属
base peak 基峰
basic lead carbonate 碱式碳酸铅
basic material 基体材料
basic metal 母材，碱性金属
basic oxide 碱性氧化物
basic reaction 碱性反应
basic salt 碱性盐，碱式盐
basic solvent 碱性溶剂
basicity 碱度
basify 碱化，使碱化
basilar 基部的，基底的
basilical 重要的
basophil 亲碱的
basophilous 喜碱性的，适碱性的
basis material 基体材料
batch annealing 分批退火，箱式炉退火
batch extraction 分批萃取，分批提取
batch furnace 分批炉，分次式熔炉

batchwise operation 分批操作
bath 镀浴
bath solution 电解液
bath voltage 槽电压
bathochromic shift 长移，红移
battery 蓄电池
battery acid 蓄电池用酸
battery charger 充电器
battery cubicle 蓄电池箱
battery inner resistance 电池内阻
battery plate 蓄电池极板
Baume degree 波美度
Baume hydrometer 波美比重计，波美计
beaded 珠状的，饰以珠的
beaded pearlite 珠状珠光体，粒状珠光体
beading 形成珠状，卷边，起泡，起泡剂
beadlike 珠状的
beaked 有喙的，鸟嘴状的
beaker 烧杯
beam 声束，光束
beamy 光亮的，放光的，辐射的
bearings 轴承
bedded 层状的
bedstead 试验装置，实验台
bedye 着色，施彩色
belling 压凸加工
belt drier 带式干燥机
belt drive 皮带传动
belt grinder 带式磨床
belt grinding 皮带研磨，砂带磨光
belt sander 带式打磨机
benched 台阶形状的
benchmark 基准
bench-scale 小型的，实验室规模的
bending 挠曲，弯曲
bending deformation 挠曲变形，弯曲变形
bending machines 弯曲机
bending strength 弯曲强度
bending stress 弯曲应力
bending test 弯曲试验
bending vibration 弯曲振动

bengal isinglass 琼脂
bent 弯曲的
bent crystal 弯晶
benthic deposit 水底沉淀物
bentonite 膨润土
benzal 亚苄
benzamide 苯酰胺
benzene 苯
benzene hydrocarbon 苯系烃
benzene ring 苯环
benzene series 苯系
benzene sulfinic acid 苯亚磺酸
benzene sulfonic acid 苯磺酸
benzene sulfonic amide 苯磺酰胺
benzidine 联苯胺
benzine 汽油，挥发油，轻质汽油
benzo- 苯并
benzoic 安息香的
benzoic acid 安息香酸，苯甲酸
benzoyl 苯（甲）酰
benzyl 苄基，苯甲基
benzylidene acetone 亚苄基丙酮，苄叉丙酮
Beta brass β黄铜
Beta rays β射线
betaine 甜菜碱
betatron 电子感应加速器，电子回旋加速器
bevel 斜角；成斜面的；使成斜角，使成斜面
beveling 磨斜棱，磨斜边，成斜角
blowhole 喷水孔，通风孔，通气孔，（气）孔，铸孔，砂［气］眼，气泡
bi- 【构词成分】双，两，二，重
biassing 斜纹；斜的，倾斜的，斜纹的；偏斜地，倾斜地，对角地
biatomic 双原子的，双酸的
biatomic acid 二元酸
biaxial 双轴的，二轴的
biaxiality 二轴性，双轴性
bibasic 二元的，二代的

bibulous 吸水的，吸水性的
bibulous paper 吸水纸，滤纸
bicarbonate 碳酸氢盐，重碳酸盐，酸式碳酸盐
bice 绿色，灰蓝色，灰蓝色颜料
bichloride 二氯化物
bichromate 重铬酸盐
bichromate titration 重铬酸盐滴定
bichrome 重铬酸盐；两色的
bicolorimeter 双色比色计
bid 投标
bifarious 双重的，二纵列的
bifluoride 氟氢化物
bifunctional 双官能团的，有两种不同功能的
bilateral 双边的，两侧的，双向作用的
bilinearity 双线性
bill of material 物料清单，材料清单
billet 方钢，坯锭
billisecond 毫微秒，十亿分之一秒
bimaleate 马来酸氢盐
bimanual 用双手的，须用两手的
bimetallic 双金属的
bimetallic corrosion 双金属腐蚀
bimodal 双峰的
bimolecular reduction 双分子还原
binary 二元的，二态的
binary acid 二元酸
binary alloy 二元合金
bind 捆绑，结合，胶合，使凝固
binder 黏合剂
binding energy 键能，结合能
binodal 双节的
binoxalate 草酸氢盐
bioaccumulation 生物积累
biochemical oxygen demand (BOD) 生化需氧量
biocide 杀虫剂
biocolloid 生物胶体
bio-contact oxidation 生物接触氧化
biodegradation 生物降解

biofilm process 生物膜法
biological aerated filter 曝气生物滤池
biological control agent 生物抑制剂
biological filter 生物滤池
biological treatment 生物处理
bio-osmosis 生物渗透
biomembrane process 生物膜法
biopolymer 生物高分子
biosorption process 生物吸附法
bio-tower 塔式生物滤池
biphase 双相，双相的
biphenyl 联苯，联二苯
bipolar electrode 双极性电极
bipolarity 双极性
bipolymer 二元共聚物，二聚物
bipyridyl 联吡啶
biradical 双游离基的，二价自由基的
bismuth nitrate 硝酸铋
bismuthous 亚铋的，含三价铋的
bistability 双稳定性，双稳，双稳态
bisulfate 硫酸氢盐
bisulfide 二硫化物
bisulfite 亚硫酸氢盐
bisulphate 重硫酸盐，硫酸氢盐，酸式硫酸盐
bisulphite 亚硫酸氢盐，酸式亚硫酸盐
bisymmetric 两侧对称的
bitartrate 酒石酸氢盐
bite （齿轮、螺钉、钳子等）咬住
bituminous paint 沥青漆，沥青涂料
biuret 缩二脲
bivalence 二价
bivalent 二价的
blaching liquor 漂白液
black conversion coating 黑色转化膜
blackening 染黑法
blacking hole 涂料孔，铸疵
blacking scab 涂料疤
blacking up 泛黑
blackwash 黑涂料；给……上黑色涂料

bladed 有叶片的
bladed structure 刃状构造，叶状组织
blanch 漂白，发白，变白；漂白，粉饰
blank 坯料，半成品
blanking 下料，下料加工
blast 喷砂器；鼓风
blast cleaning 喷砂清理
blast heating 预热送风，鼓热风
blast-furnace 高炉，鼓风炉
blazed grating 闪耀光栅
blazer 燃烧体，发焰物
bleach 漂白剂；使漂白，使变白
bleaching 脱色，漂白；漂白的
bleaching agent 漂白剂
bleaching powder 漂白粉
bleeding 溶出，渗色，底层污染
bleed 泄放孔，放出的液体（气体）；渗出，漏出，抽吸
blemish 瑕疵，污点，缺点，表面缺陷
blend 混合，掺和物；混合
blendable 可掺和的，可混合的
blended 混杂的，数种混合的
blending 混合，调配，混合物；混合的
blind 掩饰，封闭；闭塞的，封闭的，不通的，隐蔽的
blinding 使人炫目的，不清晰，模糊，填塞
blink 表面浅洼型缩孔，（表面）缩陷
blister 起泡，气孔，砂眼，起皮
blister corrosion 起泡腐蚀
blistered 起泡的
blistered casting 多孔铸件
block copolymer 块状共聚物
blockage 堵塞，堵塞
blocking 黏结，结块
blooming 阳极粉化
blow down 排污
blow hole 破孔，气泡，砂眼，通风孔
blow-moulding forming 吹制成型（吹塑）
blow-off 吹出，放气

blowdown 排污，排污水
blowdown apparatus 排污装置
blower 通风机，鼓风机
blowing agent 发泡剂
blown 吹制的，吹出的，多孔的，海绵状的
blown asphalt 氧化沥青
blue annealing 软化退火，发蓝退火
blue brittleness 蓝脆
blue dip 汞齐化
blue scale 蓝卡
blue shift 蓝（紫）移
blue shortness 青熟脆性，蓝脆性
blue vitriol 胆矾，五水硫酸铜
bluing 发蓝（钢铁化学氧化）
bluish 带蓝色的，有点蓝的
blushing 发白，白化，泛白
blunt 钝的，不锋利的，生硬的
blur 污迹，模糊不清的事物；涂污，使……模糊不清，使暗淡
blurring 模糊；模糊的
blushing （涂料）雾浊，发红，褪色
boat 蒸发皿
bobbin 线筒，缠线管，缠线板，卷线轴
bobbin oil 锭子油
bobbing 抛光，振动
body centered cubic structure 体心立方结构
body centered lattice 体心晶格，体心点阵
body wrinkle 侧壁皱纹
boehmite 一水合氧化铝
boehmite process 化学氧化处理
bogen structure 弧状构造
boiling fastness 耐煮性
boiling heat 蒸发热
boiling point 沸点
boiling water resistance 耐沸水性
boiling water sealing 热水封孔处理
boilproof 耐煮的
bolt 螺栓，螺钉
bolt pin 螺栓插销

Boltzmann distribution 玻尔兹曼分布
Boltzmann's constant 玻尔兹曼常数
bond 键
bond angle 键角
bond energy 键能
bond flux 粘接焊剂
bond length 键长
bond line 黏结层
bond moment 键矩
bond order 键级
bond polarity 键矩
bond strength 键强度
bondability 黏结性，黏合性
bonded 黏着的，被连接的
bonderizing 磷酸盐（防锈）处理，磷化处理
bonding 黏合
bonding agent 键合剂
bonding orbital 成键轨道
bonding water 结合水
bonding wire 接合线，焊线
bone dry 极干燥的，十分干燥的
bonnet 阀盖
boost 增压，推进
booster charge 再充电
booster pump 升压泵
booster stages 扩增部件
boracic 含硼的，硼的
borane 甲硼烷，硼氢化物
borate 硼酸盐
borated 含硼酸的，用硼酸处理过的
borax 硼砂
borazon 氧化硼立方晶
boric 硼的，含硼的
boric acid 硼酸
boric anhydride 硼酸酐
boride 硼化物
boriding 渗硼
boro- 【构词成分】硼
boroethane 乙硼烷
borofluoric acid 氟硼酸

borofluoride 氟硼酸盐
borohydride 氢硼化物
boron 硼
boron carbide 碳化硼
boron fiber 硼纤维
boron nitride 氮化硼
boron group 硼族
boronising 渗硼法
boronize 渗硼
boronizing 渗硼，硼化处理
borosiliconizing 渗硼硅处理
bosh 浸冷（用的）水槽，（酸洗）槽
boshing 浸水除鳞，浸水冷却
bottle 瓶子；控制，把……装入瓶中
bottom coat 底漆
boundary condition 边界条件，界面条件
boundary displacement 界壁位移
boundary domain 磁域界壁，边域
boundary effect 边界效应
boundary layer 界面层，边界层
boundary tension 界面张力
bow 弓；弯曲的
bowl 球形物，离心机转筒，反射罩
bowl mill 球磨机
box annealing 箱型退火
box carburizing 封箱渗碳，箱式渗碳法
box funace 箱式炉
brackish 含盐的，微碱的，碱化的
brackish water 苦咸水，碱性水
brackishness 微碱性
Bragg's law 布拉格定律
brale 圆锥形金刚石压头
branched 分枝的，枝状的，有枝的
branched compound 支链化合物
branched hydrocarbon 支链烃
branched polymer 分支聚合物
branner 绒布磨光轮，清净机
brash 易碎的，脆的；脆性
brashy 易碎的，脆的
brass 黄铜
brassboard 实验性的，实验的，模型的

brassinolide 黄铜质
brassy 似黄铜的，黄铜色的
brastil 压铸黄铜
bravity spray gun 重力式喷枪
braze 铜焊；用黄铜制，使成黄铜色
brazing （硬）钎焊
breadboard 案板，电路试验板；为……制作，制作模拟板
breadth 宽度，幅度，外延
break line failure 保护不良
breakability 可破碎性，易破碎性
breakage 破坏，破损，裂口，破损量
breakdown 崩溃
breakdown point 击穿点，屈服点
breakdown voltage 击穿电压
breaking 断裂
breaking load 断裂载重，断裂负荷
breaking point 断裂点
breaking quenching 辉面淬火
breaking strength 断裂强度
breaking stress 破坏应力
breaking test 破坏试验
breakthrough 突破，穿透，贯穿，渗漏，临界点，转折点
breathing 通风
bremsstrahlung 韧致辐射
brevi- 【构词成分】短
brevilineal 短形的
bridging oxygen 架桥氧，桥氧
bright chrome plating 镀亮铬
bright current density 光亮电流密度
bright dipping 出光，浸亮
bright-drawn 冷拔的，精拔的
bright etching 光亮蚀刻
bright finish 光泽处理
bright goldliquid gold bright hardening 辉面硬化
bright heat treatment 辉面热处理
bright nickel 光亮镍
bright heat treatment 光亮热处理
bright pickling 光亮侵蚀

bright plating 光亮电镀
bright smooth finish 光亮平滑镀层
brightener breakdown product 光亮剂的分解物
brightener 抛光剂，上光剂
brightening agent 光亮剂
brightness 亮度
brightness and levelling test 光亮及整平性测试
brilliance 光泽，亮度
brine 浓盐液，浓咸水
brinell hardness 布氏硬度
brink 边缘
brinish 盐水的，咸的，苦的
briny 海水的，咸的，盐水的
briquettability 压塑性，压制性
briquetting 压块，制团
briquetting press 压片机，压块机
brittle 易碎的，脆弱的，易损坏的
brittle fracture 脆性断裂
brittle material 脆性材料
brittle rupture 脆性破坏
brittleness 脆性
broaden 使扩大，使变宽；扩大，变阔，变宽，加宽
broadening 扩展，增宽
brom- 【构词成分】溴
bromate 溴酸盐；使与溴化合
bromated 含溴的，溴化的
bromating 溴化
bromic 溴的，含有溴的
bromide 溴化物
bromimetry 溴量法
bromine 溴
bromine method 溴量法
bromizate 溴化，用溴处理
bromophenol blue 溴酚蓝
bronze 青铜
bronze plating 镀青铜
bronzing 镀青铜，青铜化，着青铜色
brush coating 刷涂

brush plating	刷镀
brushed	拉丝
brushed brass	拉丝黄铜，磨砂铜
brushed metal	金属拉丝
brushing	刷光
bubble	膜泡，气泡
bubble trace on ED film	电泳气泡
buchner funnel	布氏漏斗
buff	擦光轮
buffer	缓冲剂
buffer capacity	缓冲能力，缓冲容量
buffer salt	缓冲盐
buffer solution	缓冲溶液
buffing	软轮磨光
buffing compound	抛光蜡
buffing compound removal	除蜡水
buffing machine	抛光机
buffing wheel	抛光布轮
bug	缺陷
bulge test	膨胀测试
bulging	撑压加工
bulk density	体积密度
bulk heat treatment	整体热处理
bulk phase	本体相
bunch	串，突出物；使成一串
bundle	捆，包
buret	量管
buret clamp	滴定管夹
buret clamp pincers	滴定管夹
buret stand	滴定管架
burette	滴定管
burgers vector	柏氏矢量
burn mark	糊斑
burner house	燃烧室
burning	烧伤
burning of anodic oxide coating	阳极氧化膜烧焦
burnt deposite	烧焦镀层
burr	毛边，毛刺
burring	冲缘加工
burst test	胀裂试验（水压）
bursting	膨裂
busbar	汇流排，母线
bushing	衬套，轴衬，套管
butadiene	丁二烯
butane	丁烷
butanedioic acid	丁二酸
butanediol	丁二醇
butanol	丁醇
butanone	丁酮
butene	丁烯
butenediol	丁烯二醇
butine	丁炔
butyl alcohol	丁醇
butyl formate	甲酸丁酯
butyl	丁基
butynediol	丁炔二醇
by product	副产物
bypass heat treatment	旁路热处理

C

cabinet dryer 干燥箱
cable channel 电缆槽，电缆管道
cable lacquer 电缆漆
cable oil 电缆油
cable trunk 电缆管道，电缆主干线
calculating 计算；计算的
cadmiferous 含镉的
cadmium 镉
cadmium acetate 醋酸镉
cadmium chloride 氯化镉
cadmium hydroxide 氢氧化镉
cadmium nitrate 硝酸镉
cadmium plating 镀镉
cadmium red 镉红
cadmium standard cell 镉标准电池
cadmium sulfate 硫酸镉
cadmium sulfide 硫化镉
cadmium sulphide 硫化镉
cadmium yellow 镉黄
caesious 青灰色的
cake 滤渣
cage 隔离罩，罐笼，框，罩，笼，护圈，保持架
cage construction 骨架构造[结构]，笼式结构
cage drum 笼式滚筒
cage effect 笼蔽效应
cage grid 笼形栅极
cage mill 笼式粉碎机
cage zone melting 笼式区域熔化
caking 黏结，凝结性，结饼
caky 凝固的，饼状的
calandria 加热体（管群），排管式堆容器，蒸发设备，排管，排管式，排管体
calc- 钙（盐），石灰
calcar 熔炉
calcareous 钙质的，石灰质的
calcia 氧化钙

calcic 钙的，含钙的
calciferous 含钙的，含碳酸钙的
calcinate 煅烧产物；煅烧
calcination 煅烧，焙烧，焙解
calcine 烧成石灰；煅烧
calcium 钙
calcium bleach 漂白粉
calcium carbonate 碳酸钙
calcium chloride 氯化钙
calcium hydroxide 氢氧化钙
calcium hypochlorite 次氯酸钙
calcium oxide 氧化钙
calcium phosphate 磷酸钙
calcium silicate 硅酸钙
calcium sulfate 硫酸钙
calcon 钙试剂
calculate 计算
calculated 计算的
calculation 计算
calculous 结石的，似石的，多石的
calender 压光机，压延机
calendering molding 压延成形，轮压成形
calibrate 校正，标定，分度，刻度
calibration 校准
calibration curve 标定曲线，校正曲线
calibration error 标定误差
calibration instrument 校准仪器
calibration reference 校准基准
caliper 卡规
calliper gauge 孔径规
calomel 甘汞，氯化亚汞
calomel electrode 甘汞电极
calomel electrode potential 甘汞电极电位
calomel half cell 甘汞半电池
calori- 【构词成分】热
calorescence 发光热线，灼热
caloric 热量；热量的
caloricity 生热力，热值

calorify 加热于，使热
calorite 耐热合金
calorize 以铝处理表面，渗铝
calorstat 自动调温器，节温器，恒温箱
calpis 乳浊液
camber 翘曲
camel hairbrush （涂料用的）毛刷
camphor 莰酮，樟脑
cancellation 取消，删除
cancelling 取消，消除
cancelling circuit 补偿电路
cankerous 疡的，腐蚀的，使害溃疡的
cannular 管状的，筒状的
capacitance 电容
capacitance curve of electrical double layer 电极双电层电容曲线
capacitive approach switch 电容式接近开关
capacitive character 电容性
capacitive charging capacity 电容充电容量
capacitive circuit 电容电路，容性电路
capacitive coupling 电容耦合，容性耦合
capacitive current 电容电流，残余电流
capacitive feedback 电容性反馈
capacitive filter 电容滤波器
capacitive grid current 栅极电容电流
capacitive impedance 电容性阻抗，容抗
capacitive reactance 电容性电抗，容抗
capacitor 电容器
capacitor batteries 电容电池
capacity 容量
capacity attenuation 容量衰减
capillarity 毛细管现象
capillary 毛细管；毛细管的，毛状的
capillary action 毛细管作用
capillary analysis 毛细管分析
capillary column 毛细管柱
capillary constants 毛细管常数
capillary electrometer 毛细管静电计
capillary electrophoresis 毛细管电泳法
capillary gas chromatography 毛细管气相色谱
capillary gel electrophoresis 毛细管凝胶电泳
capillary phenomenon 毛细管现象
capillary pipet 毛细吸管
capillary tube 毛细管
capillary tube method 毛细管法
capillary viscometer 毛细管黏度计
capillator 毛细管比色计
capillometer 毛细管测试仪
capric acid 癸酸
caproic acid 己酸
capryl 癸酰，辛酰基，辛基
caprylic acid 辛酸
captance 容抗
carbalkoxylation 烷氧羰基化
carbamate 氨基甲酸盐
carbamic acid 氨基甲酸，氨基甲酸酯
carbamide 脲，尿素，碳酰二胺
carbanion 碳负离子
carbide 碳化物
carbinol 甲醇
carbo- 【构词成分】碳
carboamidation 氨羰基化
carbobenzoxy chloride 苄氧甲酰氯
carbocation 碳正离子
carbocyclic compound 碳环化合物
carbohydrate 碳水化合物
carbolate 酚盐，苯酚盐
carbolic acid 苯酚
carbon 碳
carbon acid 碳酸
carbon adsorption chloroform extraction 碳吸附-氯仿萃取物
carbon arc 碳弧灯
carbon black 炭黑
carbon chain 碳链
carbon compound 碳化合物
carbon content 碳含量
carbon dioxide 二氧化碳
carbon disulfide 二硫化碳

carbon electrode 碳电极
carbon fiber 碳纤维
carbon fiber electrode 碳纤维电极
carbon filter 活性炭过滤器
carbon group 碳族
carbon monoxide 一氧化碳
carbon potential 碳势
carbon restoration 复碳
carbon steel 碳素钢
carbon tetrachloride 四氯化碳
carbon zinc batter 碳锌电池
carbon-carbon composite 碳-碳复合材料
carbon-paste electrode 碳糊电极
carbonate 碳酸盐；碳化
carbonate hardness 碳酸盐硬度
carbonation 碳酸化
carbonic 碳的，含碳的
carbonic acid 碳酸
carbonitriding 渗碳氮化法，碳氮共渗
carbonization 碳化
carbonyl compound 羰基化合物
carbonyl group 羰基
carboxyl acid 脂肪酸
carborundum 金刚砂
carboxyl 羧基
carboxylation 羧化
carboxylic acid 羧酸
carboxylic acid derivative 羧酸衍生物
carboxymethyl cellulose 羧甲基纤维素
carburization 渗碳
carburize （使）渗碳，碳化
carburized case depth 浸碳硬化深层，渗碳层厚度
carburized depth 渗碳深度
carburized structure 渗碳组织
carburizer 渗碳剂
carburizing 碳化
carburizing flame 渗碳焰
carburizing furnace 渗碳炉
carburizing reagent 渗碳剂
carburizing salt 渗碳盐

carburizing steel 渗碳钢
carcase 框架，架子
carmine 胭脂红，深［洋］红色；深［洋］红色的
carrier 载体
carrier fluid 载液
cascade 串联的
case depth 硬化深度，渗碳层深度，表面深度
case hardening 表面硬化
case quenching 表面淬火
casing 外壳，框架
CASS-test 铜加速乙酸盐雾试验
cast iron 生铁，铸铁
cast steel 生钢，铸钢
castor oil 蓖麻油
casting 熔铸法，铸（件），铸造
casting forming 浇铸成型（铸塑）
casting steel 铸钢
catabolite 降解产物，副产物
catalysis 催化
catalyst 催化剂
catalyst carrier 催化剂载体
catalyst selectivity 催化剂选择性
catalyst surface area 催化剂表面积
catalytic analysis 催化分析
catalytic current 催化电流
catalytic dehydrogenation 催化脱氢
catalytic hydrogenation 催化氢化
catalytic oxidation 催化氧化
catalytic polymerization 催化聚合
catalytic reaction 催化反应
catalytic reduction 催化还原
catalytic wave 催化波
catalyzer 催化剂
cataphoresis 电泳，电渗
cataphoretic 阳离子电泳的
catastrophic 大变动的，灾难的
catastrophic oxidation 高温剧烈氧化
catchment 集水区，集水量，汇水
cathode 阴极

cathode efficiency 阴极效率
cathode layer 阴极层
cathode polarization 阴极极化
cathode ray 阴极射线
cathode ray luminescence 阴极射线发光
cathode ray oscillograph 阴极射线示波器
cathode sputtering 阴极溅射
cathodic 阴极的，负极的
cathodic coating 阴极镀层
cathodic current 阴极电流，阴极波
cathodic current density 阴极电流密度
cathodic current efficiency 阴极电流效率
cathodic electrocleaning 阴极电除油
cathodic inhibitor 阴极抑制剂
cathodic polarization 阴极极化
cathodic protection 阴极保护
cathodic reaction 阴极反应
cathodic reduction 阴极还原
cathodic vacuum etching 阴极真空蚀刻
cathodic wave 阴极波
catholyte 阴极电解液
cation 阳离子
cation exchange 阳离子交换
cation exchange membrane 阳离子交换膜
cation exchange resin 阳离子交换树脂
cation soap 阳离子皂
cationic detergent 阳离子型洗涤剂
cationic dye 阳离子染料
cationic polymerization 阳离子聚合
cationic surface active agent 阳离子表面活性剂
cationic surfactant 阳离子表面活性剂
cationoid 类阳离子
cationotropic rearrangement 正离子转移重排
cause description 原因说明
caustic alkali 苛性碱
caustic baryta 氢氧化钡
caustic embrittlement 碱脆
caustic lime 苛性石灰
caustic potash 氢氧化钾

caustic soda 烧碱
caustic wash 碱洗
causticity 苛性度
caustification 苛化
cavitation 空穴现象，空泡作用，穴蚀
cavitation corrosion 空泡腐蚀
cavitation damage 涡穴损伤，空泡损伤
cavitation erosion 涡穴冲蚀，孔蚀，气蚀
cavity 孔穴，凹处
cell 电池
cell constant 电池常数
cell potential 电池电位，电池电势
cellophane 玻璃纸
cellosolve 溶纤剂，纤维素溶剂
cellosolve acetate 乙酸溶纤剂
cellular polystyrene 泡沫聚苯乙烯
celluloid 赛珞璐
cellulose 纤维素
cellulose acetate 醋酸纤维素，纤维素乙酸酯
cellulose acetate membrane 醋酸纤维膜
cellulose hydrate 纤维素水合物
cellulose material 纤维素材质
cellulose nitrate 硝酸纤维素，硝化纤维
celsius scale 摄氏温标
celsius thermometer 摄氏温度计
cement 水泥，接合剂；巩固，加强
cement steel 渗碳钢，表面硬化钢
cementation 渗碳
cementing material 黏合剂，胶凝材料
cementite 渗碳体
center ion 中心离子
centigrade 摄氏的
centinormal 百分之一当量浓度的
centrifugal 离心机，转筒；离心的
centrifugal analysis 离心分析
centrifugal compressor 离心压缩机
centrifugal blower 离心通[鼓]风机
centrifugal dehydrator 离心脱水机
centrifugal extractor 离心萃取器[脱水机]

centrifugal fan 离心通风机，离心式鼓风机
centrifugal force 离心力，地心引力
centrifugal pump 离心泵
centrifugal sedimentation 离心沉降
centrifugal separator 离心分离器［澄清机］
centrifugal washer 离心洗涤器
centrifugation 离心分离
centrifuge 离心机［干燥剂，分离机］
centrifuging 离心
ceramic 陶瓷
ceramic matrix composite 陶瓷基复合材料
ceramic membrane 陶瓷膜
ceria 二氧化铈
ceric oxide 二氧化铈
ceric sulfate 硫酸高铈
cerimetric titration 铈滴定
cerimetry 铈量法
cerium 铈
cerium sulphate method 硫酸铈法
cerium chloride 氯化铈
cerium hydroxide 氢氧化铈
cerium nitrate 硝酸铈
cerous 铈的，三价铈的
cerous nitrate 硝酸亚铈
cerous salt 三价铈盐
chafing 摩擦的，防擦的
chafing-fatigue 摩擦疲劳
chain hoist 手动葫芦，链式起重机
chain drive 链传动，利用链条传送动力的装备
chain transmission 链条传动
chalcanthite 胆矾，蓝矾
chalcogen 氧族，硫族元素
chalcogenide 氧族化合物，硫族化物
chalking 粉化
chamfer 倒角，斜面，凹槽；去角，挖槽，斜切
chamber controller 恒温槽控制器
chamber dryer 箱式干燥机，箱式干燥器

chamber filter press 箱式压滤机
chamfering machine 倒角机，刨边机
channel 凹槽，通道
characteristic 表征；典型的，特有的，表示特性的
characteristic absorption peak 特征吸收峰
characteristic curve 特性曲线
characteristic curve of hardening 淬火特性曲线
characteristic frequency 特征频率，固有频率
characteristic radiation 特性辐射，标识辐射
characteristic X-ray 特征X射线
characterize 表现……的特色，刻画……的性格
charge 电荷，填充，充电
charge balance 电荷平衡
charge balance equation 电荷平衡式
charge carriers 载流子（指电子和空穴）
charge characteristic 充电特性
charge conjugation 电荷共轭
charge current 充电电流
charge curve 充电曲线
charge density 电荷密度
charge efficiency 充电效率
charge modification 荷电修改
charge neutralization 荷电中和
charge sensitivity 电荷灵敏度
charge step 电量阶跃法
charge symmetry 电荷对称性
charge transfer electrode 电荷转移电极
charge transfer resistance 电荷传递电阻
charge up 充电，使绝缘物带电
charge voltage 充电电压
charged particle 带电粒子
charger 装料机，充电器
charging 装料，充电
charging apparatus 装料设备
charging current 充电电流
charging hopper 加料漏斗

Charpy test 夏比试验（单梁冲击试验）
charring 碳化
chassis black paint 底盘黑涂料
chatelier-type microscope 倒置式显微镜
chatter 颤震（机器），波纹（加工面）
chattering 颤动
check 检查，裂纹，细裂痕，校核
check analysis 检验分析
check point 查核点
check test 检查试验
checked by 初审，核对
checking 龟裂，浅裂纹，裂痕（涂料）
checking code 检码
chelant 螯合[掩蔽]剂
chelate 螯合；螯合的，有螯的
chelate compound 螯合物
chelate effect 螯合效应
chelating agent 螯合剂
chelating ion exchange resin 螯合离子交换树脂
chelation 螯合，螯环化
chelatometric titration 螯合滴定
chelatometry 螯合滴定法
chemical 化学的
chemical actinometer 化学光量计
chemical activity 化学活性
chemical affinity 化学亲和力
chemical analysis 化学分析
chemical apparatus 化学仪器
chemical atomic weight 化学原子量
chemical attack 化学侵蚀
chemical balance 化学天平
chemical bleaching 化学漂白
chemical bond 化学键
chemical brightening 化学光亮化，化学抛光
chemical cell 化学电池
chemical cleaning 化学清洗
chemical coagulation 化学凝聚
chemical composition 化学组成
chemical compound 化合物

chemical constitution 化学结构，化学组成
chemical conversion coating 化学转化膜
chemical corrosion 化学腐蚀
chemical decaling 化学除锈剂
chemical degreasing 化学脱脂法
chemical derusting 化学除锈剂
chemical dissolving box 药剂溶解箱
chemical double layer 化学双电层
chemical equilibrium 化学平衡
chemical factor 化学因数
chemical heat treatment 化学热处理
chemical ionization source 化学离子源
chemical methods for local thickness 化学法测试局部厚度法
chemical oxidation 化学氧化
chemical oxygen demand (COD) 化学耗氧量
chemical passivation 化学钝化
chemical plating 化学镀
chemical polishing 化学抛光，化学研磨
chemical precipitation 化学沉淀
chemical proportioner 比例加药器，比例加药机
chemical reaction 化学反应
chemical resistance 耐药品性
chemical reversibility 化学可逆性
chemical sewage 化学污水
chemical shift 化学位移
chemical stripping 化学溶出
chemical treatment 化学处理
chemical vapor deposition (CVD) 化学蒸镀，化学气相沉积
chemical wear 化学磨损
chemically bonded-phase chromatography 化学键合相色谱法
chemically modified electrode (CME) 化学修饰电极
chemically pure 化学纯
chemicals 化学药品
chemichromatography 化学色谱法
chemistry 化学

chemistry coarsening 化学粗化
chemometrics 化学计量学
chiletropic reaction 螯键反应
chill crack 淬裂
chill mark 淬纹
chiller 冷却器
chip 片纹；切屑
chip resistant paint 抗崩裂性
chip test 冲击试验
chipping 剥离
chipping resistance 耐崩裂性
chiral chromatography 手性色谱法
chiral molecule 手性分子
chisel 凿；雕，刻，凿
chitin 甲壳素，壳多糖
chitobiose 壳二糖
chitosan 壳聚糖，脱乙酰壳多糖
chlorate 氯酸盐
chloride 氯化物
chlorhydric acid 盐酸
chloride 氯化物
chlorine 氯
chlorine peroxide 二氧化氯，过氧化氯
chloroethylene 氯乙烯
chloroform 氯仿，三氯甲烷
chloroprene 氯丁橡胶
chlorosulfenation 氯亚磺酰化
chlorosulfonation 氯磺酰化，氯磺酰化作用
chlorosulfonic acid 氯磺酸
cholromethylation 氯甲基化
chroma 鲜艳度，饱和度，纯度
chromate 铬酸盐；用铬酸盐处理
chromaticity 色度，染色性
chromating 铬酸盐处理，钝化，染色处理
chromatographic analysis 色谱[层析]分析
chromatography 色谱法，层析法
chrome 铬
chrome heat resistance steel 铬系耐热钢
chrome-molybdenum steel 铬钼钢

chrome plating 镀铬
chrome yellow 铬黄
chromic acid 铬酸
chromic anhydride 铬酸酐
chromic chloride 氯化铬
chromic oxide 铬绿
chromic salt 三价铬盐
chromium 铬
chromium chloride 氯化铬
chromium compound 铬化合物
chromium hydroxide 氢氧化铬
chromium-nickel steel 铬镍钢
chromium-plated 镀铬的
chromium plating 镀铬
chromium steel 铬钢
chromium sulfate 硫酸铬
chromium trioxide 三氧化铬
chroming 镀铬
chromized steel 渗铬钢，镀铬钢
chromizing 渗铬
chromophore 生色团
chronoamperometry 计时安培分析法[电流法]
chronocoulometry 计时库仑法[电量法]
chronopotentiometry 计时电位滴定法
chue 色相
chuck 夹具
chute 滑道，滑槽，斜槽
cinder inclusion 夹渣，包渣
cinnamic alcohol 肉桂醇
circuit 回路
circuit breaker 断路开关，断路器
circular chromatography 环形色谱法
circular table 圆盘，旋转盘
circulation 循环
circulation pump 循环泵
circulating water 循环水
circulation valve 循环阀
cis 顺式
cis isomer 顺式异构体
citrate 柠檬酸盐

citric acid　柠檬酸，枸橼酸
clamp　夹钳；夹紧，固定住
clamping device　夹紧装置，夹持工具
clad steel　复合钢
clamp-off　铸件凹痕，掉砂
clamping fixture　胎具，夹具
clarification　澄清，净化
clarifier　沉淀池，澄清池
classical　古典的，经典的，传统的
classical method of polarography　经典极谱法
classification　分类，整理
clean up　清除
cleanability additives　清洗剂
cleaner　脱脂剂
cleaning agent　净化剂
cleaning equipment　洗涤装置
cleaning section　清洗段
cleaning solution　清洗液
cleanness　清扫
clear coating　清漆
clear lacquer　清漆
clearance　间隙，清除
cleavage　劈开，分裂，解理
cleavage fracture　解理断裂，可裂性破坏
clip　切断，夹住
clock wise　顺时针方向
clog　障碍；堵塞，阻塞
clog up air lines　堵塞空气管
close packing of spheres　球密堆积
closed loop controls　闭环控制
closed-loop regulation system　闭环调节系统
cloud point　浊点
clouds　起云
cluster　颗粒；使聚集
coacervation　凝聚
coadsorption　共吸附
coagulant　混凝剂，促凝剂
coagulation　凝结
coalescent　成膜剂

coalescence　聚结，合并
coarse crushing　粗碎
coarse grained　粗颗粒的，粗纹理的
coarse grinding　粗磨，粗粉碎
coarse pearlite　粗粒珠光体
coarsening　结晶粒粗大化，粗化
coating　涂层，覆盖层
coating adhesion test instrument　漆膜附着力试验仪
coating equipment　涂布机
coating gauge　涂层计
coating impact instrument　漆膜冲击仪
coating mass　膜质量
coating material　涂料，涂层材料，饰面材料，覆盖材料
coating surface impact instrument　漆膜表面冲击仪
coating thickness　膜厚度
coating weight　膜重，涂层重量
coaxial cable　同轴电缆
cobalt　钴，钴类颜料，由钴制的深蓝色
cobalt chloride　氯化钴，二氯化钴
cobalt compound　钴化合物
cobalt hydroxide　氢氧化钴
cobalt lithium oxide　钴酸锂
cobaltic ion　钴离子
cobwebbing　拉丝现象，蛛网，裂痕
coconut oil　椰子油
co-current regeneration　顺流再生
coefficient　系数
coefficient of cubic expansion　体积膨胀系数
coefficient of elasticity　弹性系数
coefficient of friction　摩擦系数
coefficient of linear expansion　线膨胀系数
coefficient of scatter　散射系数
coefficient of stress concentration　应力集中系数
coefficient of thermal conductivity　热导率
coefficient of thermal expansion　热胀系数
coefficient of viscosity　黏度系数

coefficient of variation	变异系数
coercive field	矫顽力
coercive force	矫顽力
cohesion	内聚力，凝聚
cohesion energy	内聚能
coil anodizing	卷材阳极氧化
coil spring	螺旋弹簧，圆簧，卷簧
coining	压印加工
coining forming	模压成型
cold chamber die casting	冷式压铸
cold crushing strength	常温抗碎强度，冷碎强度
cold crystallization	冷结晶
cold cure	冷硬化
cold drawing	冷拔
cold forging	冷锻
cold hobbing	冷挤压，冷挤压制模法
cold molding	冷模
cold pressing	冷压
cold quenching	水冷淬火
cold rolled steel sheet	冷轧钢板
cold rolled steel strip	冷轧带钢
cold rolling	冷轧，冷压延
cold shortness	低温脆性，冷脆性
cold stretch	冷拉伸，冷拉
cold treatment	冷处理
cold working	冷加工
collapse	塌陷；使倒塌
collargol	胶体银
collection efficiency	收集效率
colloid	胶体；胶质的
colloid stabilizer	胶体稳定剂
colloidal	胶状的，胶质的
colloidal electrolyte	胶态电解质
colloidal particle	胶粒
colloidal property	胶体特性
colloidal solution	胶体溶液
colloidal state	胶态
colloidal suspension	胶态悬浮液
color	彩色
colophony	松香
color bleeding	渗色
color center	色心
color chart	色卡，色板
color comparator	比色仪，比色器
color comparison tube	比色管
color difference	色差
color difference meter	色差计
color fastness	颜色牢固度，色牢度
color floating	浮色
color index	比色指数
color matching	色匹配度，调色
color mottle	色斑
color pigment	着色颜料
color plated zinc	彩锌
color saturation	色饱和度
color scheme	色彩设计
color sealer	同色封底涂层
color specification	色的规格
color standard panel	标准颜色板
color strength	着色力
color tolerance	色的许容差
colorant	着色剂
coloration	上色，着色，染色
colored passivation	彩色钝化
colorimeter	比色计，色度计
colorimetry	比色法
colorrendition	彩色再现
colour change interval	变色范围
colour tolerance	允许色差
colouring	着色；把……涂颜色
column chromatography	柱色谱法，柱层析法
columnar	柱状的，圆柱的
columnar structure	柱状结构，柱状组织
combination pH electrode	复合 pH 电极
combined coating of anodic oxide and organic films	阳极氧化涂装复合膜
combustion property tester	燃烧性能测定仪
combustion	燃烧，氧化
comemtite	渗碳体

command 指令；命令，指挥，控制
commercial 工业的，商业的
comminution 粉碎
commissioning 试运转，试车
common ion effect 同离子效应
compact 紧密层；巧便携的，压紧
compact double layer 紧密双电层
compaction 压实
compacting molding 压实成型
compactor 压缩器
compartmentalization 相迁移
compatibility 相容性
compatible 兼容性
compensating 补偿，补助，修正
compensator 补偿器
complete concent ration polarization 完全浓度极化
complex agent 络合剂
complex compound 络合物
complex ion 络离子
complex reaction 复合反应
complex salt 复盐
compleximetry 配位滴定法，络合滴定法
complexometric titration 络合滴定法
component 组分，成分
composite material 复合材料
composite plating 复合电镀，弥散电镀
composite wave 综合波
composition 组成，成分
compositional depth profile 成分深度剖析
compound corrosion test 复合腐蚀试验
compound die 复合模
compound molding 复合成型
compound 化合物
compressed air drying 压缩空气干燥
compressibility 可压缩性
compression 压缩
compression molding 压缩成型
compressing 压缩加工
compression pump 压缩机，压气机
compression ratio 压缩比
compression stress 压应力
compression test 抗压试验
compressive 压缩的，有压缩力的
compressive strength 耐压强度，抗压强度
comproportionation 归中反应，化合反应
computer simmlation 计算机模拟
concave spot 凹点
concave 凹面；成凹形；凹的
concent ration overpotential 浓度过电势
concentrate 浓缩物
concentrated 浓的
concentrated acid 浓酸
concentrated gasket 浓水隔板
concentrated nitric acid 浓硝酸
concentrated solution 浓溶液
concentrated sulfuric acid 浓硫酸
concentrate 提浓，浓缩
concentrating 浓缩物；注意，集中，聚集
concentration cell 浓差电池
concentration distribution 浓度分布
concentration gradient 浓度梯度
concentration limit 极限浓度
concentration overvoltage 浓差超电势
concentration polarization 浓差极化
concentration potential 浓差电势
concentration profile 浓度分布曲线
concentration 浓度，浓缩，集中，专心，关注
concrete 混凝土
condensate 冷凝物
condensation 缩合，冷凝
condensation polymerization 缩合聚合作用
condensation product 缩合物
condensation reaction 缩合反应
condensed phosphates 浓缩磷酸盐
condensed water 冷凝水
condenser 冷凝器
condensing agent 缩合剂
condition heat treatment 预备热处理
condition of equilibrium 平衡条件
conditional potential 条件电位

conditional stability constant 条件稳定常数
conditioning 调整
conductance water 电导水
conductance 传导性，导电度
conducting 传导
conducting paint 电导性涂料
conducting polymer 导电聚合物
conducting salt 导电盐
conduction band 导带
conduction electron 传导电子
conduction heating 导电加热
conductive rubber 导电橡胶
conductivity 电导率，导电系数，传导性
conductivity gauge 电导率仪
conductivity measuring apparatus 电导率测量装置
conductivity meter 电导仪
conductivity water 电导水，蒸馏水
conductometric analysis 电导分析法
conductometric titration 电导滴定法
conductometry 电导法
conductor 导体
cone 锥形
confidence interval 置信区间
confidence level 置信水平
configuration 构造，构型，结构，配置，外形
configurational interaction 阵列的交互作用
confining liquid 封闭液
confirmation 证实
confirmatory reaction 验证反应
conformation 构象
conformational effect 构象效应
conformational isomer 构象异构体
congelation 凝固，凝结
congruent transformation 等成分变化
conical 圆锥的，圆锥形的
conical ball mill 锥形球磨机
conical beaker 锥形瓶
conical cup test 圆锥杯突试验，锥形杯测试
conjugate 结合；使成对，使结合；共轭的，结合的
conjugate acid 共轭酸
conjugate acid base pair 共轭酸碱对
conjugate addition 共轭加成
conjugate base 共轭碱
conjugated 共轭的，成对的
conjugated diene 共轭二烯烃
conjugated double bond 共轭双键
conjugated effect 共轭效应
conjugated system 共轭体系
conjugation 共轭
connect 连接，接线
connecting block 接线板
connecting fittings 连接配件，连接件
connector plug 连接插头
connector socket 接线插座
consecutive 连贯的，连续不断的
consecutive reaction 连锁反应
conservation 清洁，保存，保持，保护
consistency 稠度，浓度，相容性
consistometer 稠度计
consolute 共溶性；共溶性的
consolute temperature 混溶温度，共溶温度
constant 常数，恒量；不变的，恒定的，经常的
constant boiling mixture 恒沸点混合物
constant current anodizing 恒定电流阳极氧化
constant current charge 恒流充电
constant current chronopotentiometry 恒电流计时电势法
constant current discharge 恒流放电
constant current electrolysis 恒电流电解法
constant force mode 恒力模式，定力模式
constant height mode 恒高模式
constant load 恒载荷
constant load test 定载试验

constant moistened sample 恒湿样品
constant phase element 常相位角元件
constant potential electrolysis 恒电压电解
constant pressure 恒压
constant rate drying 恒速干燥
constant temperature 定温，恒温
constant temperature and humidity cabinet 恒温恒湿箱
constant temperature circulator 恒温循环泵
constant voltage anodizing 恒定电压阳极氧化
constant voltage constant current charge 恒流恒压充电
constant voltage life test 恒压寿命测试
constant watt discharge 恒功率放电
constant weight 恒重
constituent 构成部分；构成的，组成的，选举的
constitution water 化合水
constitution 结构
constitutional formula 结构式
constitutional isomer 结构异构体
consume 消耗，消费，耗尽，毁灭
consumption 消耗
contact 接点；接触
contact angle 接触角
contact corrosion 接触腐蚀
contact electrode 接触电极
contact potential 接触电位
contact process 接触法
contact resistance 接触电阻
contact surface 接触表面
contactor 触头，接触器，触点，开关
container 蓄电池槽，蓄电池壳，容器
containing 包含，含有
contamination 污染，弄脏，毒害，玷污
contamination level 污染量
content 含量
continous furnace 连续炉
continuity 连续性，持续性

continuous annealing furnace 连续退火炉
continuous cooling transformation 连续冷却转变
continuous cooling transformation diagram 连续冷却转变图
continuous distillation 连续蒸馏
continuous feed 连续投药
continuous galvanizing line 连续热镀锌线
continuous ion exchange operation 连续离子交换操作
continuous linear array 连续线性，连续线性阵列
continuous operation 连续操作
continuous phase 连续相
continuous plating 连续镀
continuous rectification 连续精馏
continuous settling equipment 连续沉淀装置
continuous spectrum 连续光谱
continuous variation method 连续变化法
continuous wave NMR 连续波核磁共振
contour 轮廓
contraction 收缩
contraflow 反流
contrast 对比度，衬度
contrast test 对比试验
contrifuger 离心干燥机
control 控制
control cubicle 控制柜
control equipment 控制设备
control panel 控制面板
controllability 可控制性
controlled atmosphere heat treatment 可控气氛热处理
controlled potential coulometry 控制电位库仑法
controlled potential electrolysis 控制电位电解法
convection 对流
conventional 传统的，常见的，惯例的
conventional aeration 普通曝气

conventional heat treatment 普通热处理
conventional polarograph 直流极谱，常规极谱
conversion 转化，变换
conversion coating 转化膜
convert 转换，转变，变换
convex 凸状，凸形；凸的，凸面的
conveyer belt 输送带
conveying chain 输送链
convolution 卷积，回旋，盘旋，卷绕
convolution principle 卷积原理
convolution spectrometry 褶合光谱法
convolution transform 褶合变换
cool color 冷色
coolant 冷却液，冷却剂
cooler 冷却器
cooling 冷却
cooling bath 冷却浴
cooling curve 冷却曲线
cooling medium 冷却介质
cooling pipe 冷却管
cooling pond 冷却池
cooling rate 冷却速度
cooling schedule 冷却制度
cooling tower 冷却塔
cooling unit 强冷室
cooling water 冷却水
cooling water return 循环冷却水回水
cooling water supply 循环冷却水给水
coordinate 坐标；调整，整合；并列的，同等的
coordinate bond 配位键
coordinated water 配位水
coordination 配位，协调，调和，对等，同等
coordination atom 配位原子
coordination chemistry 配位化学
coordination compound 配位化合物
coordination formula 配价式，配位式
coordination isomer 配位异构物
coordination isomerism 配位异构

coordination lattice 配位晶格
coordination number 配位数
coordination polymer 配位聚合［高聚］物
coordination theory 配位理论
copolymer 共聚物
co-polymerization 共聚合作用
copper 铜
copper accelerated acetic acid salt spray (CASS) 铜盐加速乙酸盐雾实验
copper acetate 醋酸铜
copper chloride 氯化铜
copper cyanide 氰化铜
copper hydroxide 氢氧化铜
copper immersion 置换铜
copper nitrate 硝酸铜
copper plating 镀铜
copper powder 铜粉
copper pyrophosphate plating 焦磷酸铜电镀
copper stencil printing 铜版印花
copper sulfate 硫酸铜
coprecipitation 共沉淀
copy 样板，仿形
coreactant 共反应物，共反应剂
corner effect 锐角效应
corpuscle 微粒，微粒子
corrected 校正的，改正的
correction 修正
correlation 相关，关联，相互关系
correlation analysis 相关分析
correlation coefficient 相关系数
correlative 相关的，有相互关系的
correlative double sampling 相关双重取样法
correspondence 相当于，一致
correspondence principle 对应原理
corresponding 相应的，符合的；类似，相配
corresponding state 对应状态
corrodent 腐蚀剂；腐蚀的

corroding agent	腐蚀剂
corrodkote test	腐蚀膏试验
corrosion	腐蚀
corrosion control	腐蚀控制
corrosion current	腐蚀电流
corrosion current density	腐蚀电流密度
corrosion-erosion	腐蚀-磨蚀
corrosion fatigue	腐蚀疲劳
corrosion fatigue cracking	腐蚀疲劳开裂
corrosion fatigue limit	腐蚀疲劳极限
corrosion inhibitive pigment	腐蚀抑制性颜料
corrosion inhibitor	防腐剂
corrosion penetration rate (CPR)	腐蚀速率
corrosion product	腐蚀产物
corrosion protection	防腐
corrosion rate	腐蚀速率
corrosion resistance	耐蚀性
corrosion resistant coating	防锈涂料
corrosion site	腐蚀点
corrosion test	腐蚀试验
corrosion wear	腐蚀性磨损
corrosion weight gain	腐蚀增重
corrosion weight loss	腐蚀失重
corrosive	腐蚀性的，腐蚀性物品
corrosive wear	腐蚀损耗
corrugated board	瓦通纸，瓦楞纸板
corundum	刚玉
cosmetic defect	外观不良，外观缺陷
cosolvent	助溶剂
co-sputtering system	共同溅镀系统
coulomb	库仑
coulomb efficiency test	库仑效率试验
coulomb efficiency	库仑效率
coulomb meter	电量计，库仑计
coulombic force	库仑力
coulometric analysis	库仑分析法
coulometric titration	库仑滴定法，电量滴定
coulometry	电量分析
coulostatic	恒电量
coulostatic analysis	静电分析
coulostatic pulse	库仑脉冲法
coumarin	香豆素
counter clock wise	逆时针方向
covalent bond	共价键
countercurrent	逆流
countercurrent regeneration	对［逆］流再生
countercurrent rinsing	逆流漂洗
counter electrode (CE)	辅助电极
counter extraction	反萃取
counter flow	逆流，对流，回流，反流
counter flow cooling tower	逆流式冷却塔
counter-ion	对离子，反离子
counter-ion transport	反离子迁移
couple action	电偶作用
coupling agent	偶联剂
coupling loss	耦合损失
coupling medium	耦合介质
coupling reaction	偶联反应
coupon	试验样板，挂片
covalence	共价
covalent bond	共价键
covalent crystal	共价晶体
covalent radius	共价键半径
covalent-network solids	原子晶体
coverage	覆盖面
covering power	覆盖能力，遮盖力
crack	裂痕
cracked	裂化的
cracking	破裂，裂纹，裂化，裂解
cracking of anodic oxide coating	阳极氧化膜破坏
crackle	碎纹
cracks	（使）破裂，裂纹，（使）爆裂
craging	龟裂
cranking current	启动电流
crape masking	纸质遮盖用胶带
crater depth	弧坑深度
crater edge effect	弧坑边缘效应

cratering 电泳缩孔，涂料缩孔，凹坑
crawling 缩孔鱼眼
craze resistance additive 抗裂纹剂
crazing 细裂纹，银纹，龟裂，网印
cream 乳油
cream of tartar 酒石酸氢钾
creep 蠕变
creeping 蠕变
creeping discharge 蠕变放电
creeping wave 爬波，蠕变波
crevice corrosion 隙间腐蚀，裂隙腐蚀
Cr-free passivation treatment 无铬钝化处理
crimping 卷边
crimping tools 卷边工具
crinkling 起皱
critical 临界的
critical constant 临界常数
critical cooling rate 临界冷却速度
critical current density 临界电流密度
critical defect 极严重缺陷
critical diameter 临界直径
critical dimension 临界尺度
critical dimension loss 临界尺寸损失
critical exponent 临界指数
critical humidity 临界湿度
critical micelle concentration 临界胶束浓度
critical point 临界点
critical pressure 临界压力
critical resolved shear stress 临界分切应力
critical state 临界态
critical temperature 临界温度
critical velocity 临界速度
critical voltage 临界电压
cross aldol condensation 交叉羟醛缩合
cross bend test 横向弯曲试验
cross cut 交叉横割
cross-drilled hole 横孔
cross link 交联
crosslinked polymer 交联聚合物
crosslinked structure 交联结构
crosslinking 交联
crosslinking agent 胶联剂
cross section 横截面
cross sectioning 横向切割
crotonaldehyde 巴豆醛
crotonic acid 巴豆酸
crown ether 冠醚
crowning 凸面加工
crucible 坩埚
crucible furnace 坩埚炉
crucible tong 坩埚钳
crude copper 粗铜
crude petroleum 原油
crude rubber 生橡胶
crusher 破碎机，压碎机
crushing 压碎
crushing strength 抗碎强度
cryatallization 结晶，结晶化
cryochem process 低温化学法
cryogen 冷却剂
cryogenic temperature 制冷温度
cryogenic treatment 深冷处理
cryolite 冰晶石
cryopreservation 低温贮藏
cryoscopic method 冰点法
cryostat 低温恒温器
cryptometer 遮盖力计
crystal 晶体
crystal axis 晶轴
crystal cell 晶胞
crystal chemistry 晶体化学
crystal field theory 晶体场理论
crystal form 晶形
crystal glass 结晶玻璃
crystal growth 晶体生长
crystal lattice 晶格
crystal nucleus 晶核
crystal orientation 晶体方位；晶体取向
crystal structure 晶体结构
crystal structure analysis 晶体结构分析

crystal system 晶系
crystal twinning 双晶
crystal violet 结晶紫
crystal water 结晶水
crystalline 结晶，晶态
crystalline granule 晶粒状块
crystalline polymer 晶体状聚合物
crystalline precipitate 晶状沉淀
crystalline state 结晶状态
crystallinity 结晶度
crystallite 微晶
crystallizability 可结晶性
crystallization 晶化，结晶作用
crystallization point 结晶点
crystallization velocity 结晶速度
crystallize 使结晶
crystallizer 结晶器
crystallogram 结晶衍射图
crystallographic axis 晶轴
crystallography 结晶学
crystallology 晶体学
crystals distortion 晶格畸变
cubic centimeter 立方厘米
cubic meter 立方米
cubic system 等轴晶系
cullet 碎玻璃
cumulative percent 累积百分比
cup flow test 杯模式流动度试验
cup of spray gun 涂料杯，杯喷枪
cupric chloride 氯化铜
cupric cyanide 氰化铜
cupric nitrate 硝酸铜
cupric oxide 氧化铜
cupric salt 正铜盐
cupric 二价铜的
cuprite 赤铜矿
cupronickel 白铜
cuprous chloride 氯化亚铜
cuprous hydroxide 氢氧化亚铜
cuprous iodide 碘化亚铜
cuprous oxide 氧化亚铜

cuprous salt 亚铜盐
curie temperature 居里温度
curing 硫化
curing agent 固化剂
curing by polymerization 聚合干燥
curing catalyst 固化促进剂
curing time 硬化时间
curl bending 卷边弯曲加工
curling 卷曲，卷曲加工
current 电流
current attenuation 电流衰减
current density 电流密度
current density range 电流密度范围
current distribution 电流分布
current efficiency 电流效率
current flow method 通电法，电流法
current induction method 电流感应法
current integration 电流积分
current interrupt 电流中断
current magnetization method 电流磁化法
current-overpotential equation 电流-过电势公式
current response 电流响应
current reversal chronopotentiometry 电流反向计时电势法
current step 电流阶跃法
current-potential curve 电流-电位曲线
current waveform 整流波形
cursor 游标
curtain flow coater 淋幕式平面涂装机
curve 曲线
cushion 缓冲
cut off 切开，切断
cutter 切工，切刀，雕工
cutting process 切削加工
cut-off 定点，取舍点，分离点
cut-off voltage 终止电压
cutters 刀具
cutting 切削加工
cutting die 冲裁模
cutting fluid 切削液

cutting jip　切削夹具
cutting-off machines　切断机
cutting tool　切削工具
cyamelide　氰白
cyanalcohol　氰醇
cyanamide　氨基氰
cyanate　氰酸盐
cyanic acid　氰酸
cyanide　氰化物
cyanide copper plating　氰化镀铜
cyanide process　氰化法
cyanide zinc plating　氰化镀锌
cyaniding　氰化
cyanogen　氰
cyanogen chloride　氯化氰
cyanogen fluoride　氟化氰
cyanogen halogenide　卤化氰
cyanogen iodide　碘化氰
cyanohydrin　氰醇
cyclane　环烷烃
cycle life　循环寿命
cycle life test　循环寿命测试
cycles of concentration　浓缩倍数
cyclic chronopotentiometry　循环计时电势法
cyclic compound　环状化合物
cyclic ester　环酯
cyclic fatty acid　环状脂肪酸
cyclic hydrocarbon　环状烃
cyclic imide　环状亚胺
cyclic ketone　环酮
cyclic process　循环过程
cyclic voltammetry（CV）　循环伏安法
cyclic voltammetry at high scan rate　快速扫描循环伏安法
cyclic voltammetry linear scan　线性波循环伏安法
cyclic voltammetry staircase　阶梯波循环伏安法
cycling　循环
cyclization　环化
cyclobutadiene　环丁二烯
cyclobutane　环丁烷
cyclobutene　环丁烯
cycloheptane　环庚烷
cycloheptene　环庚烯
cyclohexadiene　环己二烯
cyclohexane　环己烷
cyclohexanol　环己醇
cyclohexanone　环己酮
cyclohexene　环戊烯
cyclone　旋风器
cycloolefine　环烯
cycloparaffin　环烷烃
cyclopentadiene　环戊二烯
cyclopentane　环戊烷
cyclopentene　环戊烯
cyclopolymerization　环聚合
cyclopropane　环丙烷
cyclorubber　环化橡胶
cylinder　汽缸套，圆筒，量筒
cylinder dryer　转筒干燥器
cylinder head　汽缸盖
cylinder oil　汽缸机油
cylindrical mirror analyser　筒镜型能量分析器
cystine　胱氨酸

D

dactylite 指形晶
D. C. diode sputtering system 直流二极管溅镀系统
D. C. parameter test system 直流参数测试系统
D. C. polarography 直流极谱法
D. C. power equipment 直流电源
damage 损坏；损害，毁坏
damageable 易损害的，易坏的
damaging 有破坏性的，损害的
damnify 损害，损伤
damp 潮湿，湿气；潮湿的；使潮湿
damper 阻尼器，减震器，潮湿器
damping 阻尼的，减幅的，衰减的，缓冲的
dampish 含湿气的，微湿的
dampness 潮湿，湿气，含水量
damproof 防潮湿，抗潮湿
damproof insulation 隔潮
dap 挖槽，刻痕
dapple 斑纹，花马；有斑纹的；使有斑点
darapskite 钠硝矾，硫酸钠硝石
dark 深色的
dark far infrared oven 暗远红外线烘干室
dark field microscopy 暗场显微法
dark noise 暗噪声
dark trace 暗痕
darken 使变暗，使模糊，变黑，变得模糊
darkish 浅黑的，微暗的
darkle 变暗，呈现黑色
dart 投掷，投射；向前冲
dart-drop testing 双轴弯曲试验
dart drop impact test 落锤冲击试验
dash 冲撞；猛冲，撞击，使……破灭，猛撞
dash current 冲击电流
dashed area 阴影部分

dashed line 虚线，短划线
dashout 删去，除掉
data 数据
data logging 数据记录
dated 陈旧的，过时的，有日期的
datum 数据，资料，基准点，读数起点
datum mark 基准点，标高，水准点
daub 涂抹，抹胶，打底色；底涂
daub dyeing 涂布染色
dauby 乱涂的，黏性的
daze 使眼花缭乱，使晕眩
dazedly 头昏眼花地，眼花缭乱地
dazzle 使……目眩，使……眼花
de- 【构词成分】去，消，除，减，脱，分，离，解，反，非
deacidize 脱氧，还原
deactivate 去活化，钝化，减活，使不活动
deactivation 钝化（作用），惰性化，非活动化，去活化
deactivator 去活化剂
dead soft annealing 极软退火
dead soft steel 极软钢
dead stop titration 永停滴定法
dead spring 失效弹簧
dead steel 软钢，全镇静钢，低碳钢
deadbeat 非周期的，无振荡的，无调谐的
deaden 使减弱，缓和；减弱，衰减
deadening 隔音材料，消音材料，减弱，失去光泽的材料
deaerate 使除去空气，从液体中除去气泡
deaeration 脱气
deaerator 脱气剂
dealkalization 脱碱
dealloying 脱合金元素作用
dealuminization 脱铝
deamination 脱氨基
deamplification 衰减信号

deaphaneity 透明度，透明性，透明
deaquation 脱气作用
deasil 顺时针方向地
debased 低质量的，减色的；使……减色
deboration 脱硼作用
debunching 散束，电子离散
deburr 去毛刺，倒角；清理毛［芒］刺，除去脏物
Debye ring 德拜晶体衍射图
Debye-Scherrer method （X 射线检验）粉末照相法
dec- 【构词成分】十（进的），癸
decadal 十的，由十个组成的
decagon 十角形，十边形
decagonal 十边形的
decahedral 十面体的，有十面的
decahedron 十面体
decal 印花
decalescence （钢条）吸热（变暗），因吸热加快而温度降低，相变吸热
decalescent （钢条）吸热的
decaliter 十升
decameter 十米，十公尺
decane 癸烷
decanedioic acid 癸二酸
decanoic acid 癸酸
decanone 癸酮
decanormal 十当量的（溶液）
decant 移入其他容器，轻轻倒出，用沉淀法分离
decantate 倾注洗涤
decantation 缓倾（法），倾析（法），沉淀分取（法），沉淀池
decanter 倾析器，倾注洗涤器
decarbidize 脱碳沉积，脱（焦）碳
decarbonate 除去二氧化碳，脱去碳酸
decarbonization 脱碳（作用），除碳（法），去碳，减少水中的碳酸盐
decarbonize 脱碳，除去碳（素）
decarbonizer 脱碳剂
decarbonylation 脱羰

decarboxamidation 脱酰胺
decarboxylation 去羧基作用，脱羧
decarboxylative nitration 脱羧硝化
decarburization 脱碳处理
decarburize 钢的脱碳层深度，除碳素；将某物中之碳素除去
decarburized depth 脱碳深度
decarburized layer 脱碳层
decarburized structure 脱碳组织
decarburizing 脱碳退火
decationize 去阳离子，除去阳离子
decatize 汽蒸
decay 衰变，衰减；使腐烂，使腐败，使衰退，使衰落，衰退，衰减；腐烂，腐朽
decay coefficient 衰减系数
decay rate 蜕变率，下降速度
decelerate 使减速，减速，降低速度
deceleration mode 减速模式
decelerative 制动的，减速的
decem- 【构词成分】十
decene 癸烯
decentralize 分散；使分散，疏散，划分，配置
decentralized control 局部（分散）控制
decentralized data processing 分散数据处理
decentration 偏心，不共心，除去中心化
dechlorination equipment 除氯设备
dechromization 除铬，去铬
deci- 【构词成分】十分之一
deciduation 脱落，蜕落
deciduous 脱落性的，非永久性的
decimeter 分米
decimate 十中抽一，取十分之一
decimus 十分之一的，第十
decker 层，有……层的东西，脱水机，（圆网）浓缩机，装饰者
declension 倾斜，偏差，衰微
declinate 下倾，磁差角；下倾的
declination 倾斜，偏差，衰微
decline 下降，衰退，斜面；下降，衰落

declivitous 相当陡的，向下倾斜的
declivity 下坡，倾斜
declivous 倾斜的，下坡的
decoat 去膜，除去涂层
decohere 散屑（使检波器恢复常态）
decohesion 检波器恢复常态，减聚力，解黏聚，溶散
decoil 展开（卷料）
decoke 脱碳，脱去……的碳，去焦炭，除焦
decollate 区分，分割，分开，拆散
Decolorite 多孔阴离子交换树脂
decolour 漂白，使褪色
decolourant 脱色剂的，褪色剂的，漂白剂的
decolourization 脱色，消色，褪色，漂白，去色（作用）
decolourize 使脱色，使褪色，漂白
decompacting 松散
decomposability 分解性能，可分解性
decomposable 可分解的
decompose 分解，腐烂；分解，使腐烂
decomposed 分解的
decomposite 再混合物；再混合的，与混合物混合的
decomposition 分解，腐烂，变质
decomposition voltage 分解电压
decompound 再混合物；再混合的；分解，再混合
decompress 使减压，使解除压力，减压
decompression 减压，降压，分解
decompressor 减压器，减压装置
deconcentrate 使……分散，分散
decontaminant 净化剂，去污剂，纯化剂
decontamination 净化，排除污染
decontaminate 净化，给……去污
decontamination 净化，排除污染
decorate 装饰，布置，染色，油漆，施彩
decorating 装饰，美化
decoration 装饰，装潢，装饰品
decorative 装饰性的，装潢用的

decorative chromium 装饰铬
decorative coating 装饰护膜
decorative finish 装饰性镀层
decorative intermediate coating 装饰中间层
decouple 去耦；减弱震波，消除……之间的相互影响
decrease 减少，减少量；减少，减小
decreasing function 递减函数，下降函数，单调非增函数
decrement 渐减，减缩，衰减率
decrepitation 爆裂作用，烧爆声
decrescence 减弱，减小，减退，衰减
decrescent 变小的，渐减的，亏缺的
decrustation 脱皮，出去沉淀物，表面净化
decuple 十倍；十倍的；十倍于
decurtation 缩短，切短
decussation X形交叉，十字交叉
decyanation 脱氰，脱氰基
decyl 癸基的
deduce 推论，推断，演绎出
deduct 扣除，减去，演绎
deduction 扣除，减除，推论，减除额
deduction solution 缺位固溶体
deductive 演绎的，推论的，推断的，减去的，扣去的
dedust 除尘，除灰，脱尘
dedusting 除尘，除灰
de-embrittlement 除氢
de-emulsification 解乳化
deep catalytic cracking 催化裂解，深度催化裂化
deep cycle endurance 重负荷循环寿命，重复合寿命
deep hardening steel 深硬化钢
deface 损伤外观，丑化
defat 使脱脂，除油
defeature 损坏外形；使……变形
defecate 澄清，除去污物，净化，滤净
defecation 澄清（作用），去污，净化，

提净

defecator 过滤装置，澄清器
defect 瑕疵，缺陷
defect detector 故障检测器，探伤仪
defect detecting test 缺陷检查，探伤检查
defect detection sensitivity 缺陷探测灵敏度，缺陷检出灵敏度
defect echo 缺陷回波，探伤回波
defect lattice 缺陷格子
defect resolution 缺陷分辨力
defect sanding 局部打磨
defect structure 缺陷结构，缺陷组织
defective 有缺陷的，不完美的
defective coupling 不良耦合
defective insulation 绝缘不良
defective products 不良品
defectogram 探伤图，缺陷图
defectoscope 探伤仪，探伤器缺陷检查仪
defectoscopy 探伤法，探伤检验，缺陷（尺寸）测量术
deficiency 缺陷，缺点，缺乏，不足的数额
deficient 不足的，有缺陷的，不充分的
deficit 不足（额），亏损，欠缺，短缺
defilade 遮蔽物，障碍物；遮蔽
defile 污损，弄脏，染污
defilement 污秽，弄脏
definite 一定的，确切的
deflagrability 爆燃性，易燃性，暴燃性
deflagrable 可爆燃的
deflagrate 使……爆燃，使……突然燃烧，迅速燃烧
deflashing 去毛边
deflatable 可放气的，可紧缩的
deflation 放气，抽出空气，风蚀
deflection 挠曲量，挠度，弯度
deflective 偏斜的，偏离的
deflectivity 可弯性，偏向
deflectoscope 挠度计，缺陷检查仪
deflegmate 分凝，分缩，分馏
defloculant 防絮凝剂，胶体稳定剂，悬浮剂

deflocculate 抗絮凝，反团聚
defloculation 反凝，反团聚
defoam 消泡，去泡沫
defoamer 消泡剂
defoaming agents 去泡剂，消泡剂
deform 畸形的；变形；使变形
deformability 变形度，可变形性，加工性，形变能力
deformable particles 可变形粒子
deformation 变形，畸形，失真，损伤
deformation analysis 形态分析
deformation eutectic 变形共熔物
deformation mechanism 变形历程，变形机制
deformation properties 变形性质
deformation rate 变形率
deformation stress dependencies 随应力变形
deformation structure 变形组织
deformation temperature 变形温度
deformation texture 变形织构
deformation time dependencies 随时间变形
deformation twin 变形双晶
deformation velocity 变形速度
deformation zinc alloy 变形锌合金
deformative 使变形的，引起畸形的，使……形状损坏的
deformed 畸形的，变了形的
deformed bar 变形钢筋
deformograph 形变图
defroster 除霜器，防结冰装置
defrother 消沫器，消沫剂，消泡剂
degas 脱气，除气
degasification 脱气（作用），除气
degasify 给……排气
degassing 放气
degassing flux 脱气熔剂
degassing rate 去气率
degauss 去磁，消磁
degenerate 退化的；退化

degenerative 退化的，变质的
deghdrogenation 脱氢
degradation 裂解，退化，降解
degradation testing 老化实验
degrade 降级
degreaser 去油剂，脱脂剂
degreasing 脱脂
degree of branching 分支度
degree of crosslinking 交联度
degree of crystallinity 结晶度
degree of cure 硬化度
degree of freedom 自由度
degree of hydration 水合度
degree of ionization 电离度
degree of orientation 定向度
degree of polymerization 聚合度
degree of saturation 饱和度
degree of substitution 取代度
degression 递减，下降，渐减
degressive 递减的，下降的
degum 使脱胶，使去胶
dehalogenation 脱卤作用
dehardening 减硬
dehumidification 除去湿气，除湿，空气减湿
dehumidifier 减湿剂，干燥器
dehumidify 除湿，使干燥
dehumidizer 除湿剂
dehydrant 脱水剂
dehydrate 使……脱水，脱水
dehydrated alcohol 脱水酒精，无水乙醇
dehydration 脱水，干燥，去湿，皱缩
dehydrator 脱水机，脱水剂
dehydrofreezing 脱水冷冻，脱水冷冻法
dehydrofrozen 脱水冷冻的
dehydrogenate 脱氢
dehydrogenation 脱氢（作用）
dehydrogenation annealing 脱氢退火
dehydrohalogenation 脱卤化氢
dehydrolysis 脱水作用
dehydrolyze 脱水

dehydrolyzing agent 脱水剂，反水解剂
dehydroxylation 脱羟基作用
deion 消去离子；消去电离
deionization 去离子作用，除盐，消电离作用
deionize 除去离子，消电离
deionized water 去离子水
deionized water generator 纯水装置
deionized water rinsing 去离子水洗，纯水洗
deionizer 去离子剂，脱离子剂
dejacketer 脱皮装置，除去外壳的装置
delaminate 脱层，分层，分成细层，裂为薄层，层离，剥离
delamination 起鳞，分成细层
delay 延期，耽搁
delayed fracture 延迟破断，滞后断裂
delete 删除
deletion 删除，缺失
delicacy 精密，灵敏，周到
delicate 微妙的，精美的，雅致的，柔和的，易碎的，纤弱
delimitation 定界，限定
delineate 描绘，描写，画……的轮廓
deliquesce 液化，溶解，潮解
deliquescence 潮解，溶解，液化
deliquescent 溶解的，易潮解的
deliquescent chemical 潮解剂
deliver 传送，提供，放出，履行，实现
delivery 交货
delivery deadline 交货期
delivery end 卸料，输出端
delivery gate 出水口
delocalization 离域作用，移位，不受（地域）限制
delocalize 使……离开原位，使……不受位置限制
delocalized 非定域化的，不受位置限制的
delomorphic 显形的，显著的
deltoid 三角形的
deluge collection pond 集水池，蓄水池

delusterant 消光剂，退光剂
delustering 褪光，消光
delustre 褪光，除去光泽
deluxe 高级的，豪华的，奢华的；豪华地
demagnetization 去磁，消磁，退磁
demagnetizer 退磁装置，退磁器
demarcate 划分界线，区别
demask 解蔽，暴露
demasking 解蔽（作用）
demerization 二聚作用
demetallization 脱金属
demethylate 脱甲基
demi- 【构词成分】半，部分
demineralization 去矿化物质，除盐
demineralize 去除矿物质
demineralized water 脱盐水
demineralizer 脱盐装置
demisemi 四分之一的
demist 除雾，擦去……上的雾水
demix 分层，分开，反混合（指混合液分为两层）
demolish 拆除，破坏，毁坏，推翻，驳倒
demolization 过热分散（作用）
demould 脱模
demount 卸下，拆卸，把……卸下
demountable 可拆卸的，可分离的
demulcent 缓和剂；缓和的
demulsibility 反乳化率，抗乳化性，破乳化性
demulsifiable 反乳化的，破乳化的
demulsifier 破乳剂
demulsify 反乳化，脱乳，抗乳化
demulitiplication 倍［缩，递］减
denaturant 变性剂
denaturation 变性（作用），变质
denature 改变……的性质；变性
dendriform 树状的
dendrite 树枝状结晶，枝晶，树状突
dendritic 树枝状的，树状的
dendritic crystal 枝晶，枝晶体，枝状晶体
dendritic powder 树枝状粉末

dendritic segregation 树枝状偏析
dendritic structure 树枝状组织
dendroid 树状的，分枝状的
denitration 脱硝，脱硝酸盐作用
denitridation 脱氮化层
denitride 脱氮
denitriding 脱氮
denitrification 反硝化，脱氮
denitrifier 脱氮剂
denitrify 脱氮，除去氮素
denitrogenation 除氮法，脱氮作用
denoise 降噪，消除干扰
denotative 外延的，指示的
dense 稠密的，密集的，厚的
dense structure 致密结构，密实结构
densification 密实化，封严，增浓，增稠
densifier 增密器，浓缩机，增稠剂
densify 致密，使增加密度
densimeter 密度计，比重计
densitometer 比重计，浓度计，光密度计
densitometric 密度计的
densitometry 测（光）密度术，显微测密术
density 密度
density crystallinity 密度结晶
density latitude 灰度范围
density measurement 密度测量
density method 密度法
density ratio 密度比
densograph 黑度曲线，密度曲线自动描绘仪
densometer 密度计，透气度测定仪
dent 压痕，削弱，减少；削弱，使产生凹痕；产生凹陷
denticular 小齿状的
dentoid 齿状的
denudation 剥［磨，溶，侵］蚀作用，剥蚀
denude 剥夺，使裸露
deodorant 除臭剂；除臭的，防臭的

deodoriferant 除臭剂
deoxidant 脱氧剂，还原剂
deoxidation 脱氧，去氧
deoxidizer 脱氧剂，还原剂
deoxidizing 去氧，脱氧
deoxidizing agent 脱氧剂
deoxy 除氧的，减氧的
deoxygenate 除去……的氧气，出去……中的游离氧
deoxygenation 除氧
deozonize 除臭氧，去臭氧
deozonization 去臭氧（作用）
departure 离开，脱离，变更，偏差，漂移，横距
departure angle 偏转角，倾斜角
dependability 可靠性，可信任，强度，坚固度
dependency 从属，相关（性），关系，从属物
dependent 依靠的，从属的，取决于……的，非独立的，悬挂的，悬垂的
dependent 依靠的，从
dephosphorization 脱磷
department 部门
dephlegmate 使分馏，使分凝，除去……的过量水分
dephlegmator 精馏器，分凝器
dephosphorization 除磷，去磷
dephosphorize 除［去，脱］磷
deplate 除镀层，退镀
deplete 耗尽，用尽，使衰竭，使空虚
depleted 耗尽的，废弃的
depleted electrolyte 废弃［用过的］电解液
depletion 消耗，损耗
depletion effect 贫乏效应
depletion layer 耗尽层，过渡层，阻挡层
depolarization 去极化，消磁
depolarize 退极，去极化
depolarizer 去极化剂
depolarizing agent 去极化剂

depolimerization 解聚（合）（作用）
depolimerize 使……高分子解聚合，去聚合化
depolished glass 磨砂玻璃
deposit 沉淀物；使沉积，沉淀
deposit control agent 沉淀物控制剂
deposit corrosion 沉积物腐蚀
deposite attack 沉积侵蚀
deposite lattice 沉积层点阵
deposite protection 沉积防蚀
deposited metal 沉积金属，熔覆金属
depositing reservoir 澄清池
deposition 沉［淀，堆，淤］积，附着，放置，注入，析出，喷［蒸］镀，覆盖，沉降，热离解，脱溶
deposition potential 沉积电位，析出电位
deposition rate 沉积速度
depreciate 折旧，磨损，磨耗
depressant 抑制剂，抑浮剂
depressed 压下［低，平］的，降低［减压，凹下，抑制］的
depression 降低［落］，下降，减少［弱，低，压］，衰减，弱化，抽空，排气，真空（度）
depression tank 真空箱，减压箱
depropagation 链断裂作用
depth 深度
depth of chill 冷硬深度
depth of corrosion 腐蚀深度
depth of hardening 硬化深层，淬火深度
depth profiling 深度剖析
depth resolution 深度分辨率
depurant 净化器，净化剂，纯化，净化
depurate 使净化，提纯；净化，清洁
depurative 净化剂，纯化剂；净化的，纯化的
depurator 净化器，净化剂
derate 减免，下降，降低，减少［载，额，税］
derby 金属块，粗锭，块状金属，帽状物体

dereflection 去反射
derelict 被抛弃的（东西），残留物，残余物
dereliction 抛弃物，遗弃
derepression 脱［除，消］抑制（作用）
derestrict 取消对……的限制
derivant 衍生物，衍化物
derivate 导数，派生的事物；引出的，系出的
derivation 导出，诱导，分支，引水道，偏差，衍生物，微商
derivative 衍生物，导数；派生的，引出的
derivative technique 微分技术
derivatization 衍生，衍生化，衍生化作用
derogate 减损，贬损；毁损，贬低
derusting machine 除锈机
derusting 除锈
derustit 电化学除锈法
desalinate 除去……中的盐分，使……脱盐
desalination 脱盐作用，减少盐分
desalt 除去盐分
desalter 脱盐设备，脱盐剂
desaltification 脱盐（作用）
desaminase 脱氨（基）酶，去氨基（作用）
desaturate 减小饱和度，冲淡，稀释
desaturation 减饱和，去饱和作用，稀释，冲淡颜色
desaturator 干燥剂，干燥器，吸潮器，稀释剂
descale 除去锈皮，缩小比例，降级
descaler 除磷机，氧化皮消除
descaling 除去锈垢，除锈，去锈
description 说明，叙述，描述
deselenization 脱硒
desensitisation 退敏感作用，降低灵敏度
desensitis(z)e 减少感光度，使完全不感光，钝化，使感觉迟钝
desensitizer 减感剂，退敏剂

desensitivity 脱敏性
deshielding 去屏蔽，去遮蔽
desiccant 干燥剂；去湿的，使干燥的
desiccate 使干燥，除湿，脱水；干储，变干
desiccating agent 干燥剂
desiccation 干燥，除湿，脱水，干缩，烘干
desiccation fissure 干缩裂缝
desiccator 干燥器，干燥剂，除湿器
design 设计，方案
design note 设计注释
design stress 设计应力
designability 设计性，结构性，可设计性
designable 可被区分的，可被识别的
designate 指定，指派，标出，把……定名为；指定的，选定的
designation 指定［示，明］，名称，命名，符号（表示），牌号，表示方法，标识［志，记］
designator 指定者，选择器，指示器
desilicate 脱硅，除硅酸盐
desilicification 脱硅（作用，过程），除硅酸盐
desilicify 脱硅，除硅
desilt 挖除淤泥，清淤
desilter 沉淀池，除泥池，沉沙池
desilting basin 沉淀池
desintegrate 分裂，裂变，破坏
desintegration 机械破坏，蜕变，分解，裂变物，粉碎，去整合（作用）
desintegrator 粉碎机
de-sintering 清理
deslagging 除渣，脱渣，排渣，到渣
deslicking 防滑
deslicking treatment 防滑处理
desludge 清除泥渣，消除油泥
desmic 连锁的
desmodur 聚氨基甲酸酯类黏合剂
desmolysis 解链作用，碳链分解作用
desmotrope 稳变异构物，稳变异构体

desmotropic 稳变异构的
desmotropism 稳变异构（现象）
desmut 剥黑膜
desmutting 除灰，除污
desorb 释出被吸收之物，使解除吸附
desorption 解吸，去吸附
desoxidant 脱氧剂
desoxidation 脱氧，还原
desoxidizer 脱氧剂
desoxy- 【构词成分】脱氧，去氧
desoxydate 除去臭氧
desoxygenation 脱氧作用
despiker 削峰器，峰尖校平设备
despiker circuit （脉冲）削峰电路
despiking 脉冲钝化，削峰
despin 降低转速，停止旋转，反旋转，消旋
despiralization 解旋，螺旋消失，螺旋解体
despumate 除去……的泡沫［浮渣，浮垢，杂质］，撇去（液体的）漂浮物，除去杂质
dessicant 干燥剂，防潮剂
dessicate 干燥
destabilization 不稳定，去稳定（作用）
destabilize 使动摇，使不稳定
destain （为显微镜观察）把（标本）退［脱］色
destaticizer 去静电器，脱静电剂
destearinization 去硬酯
destitution 穷困，缺乏
destrengthening 强度消失，软化
destress 放松应力
destroy 破坏，消灭，毁坏，使无效
destructibility 可破坏性，易毁坏性
destruction 破坏，毁灭，摧毁
destructional 破坏作用造成的，破坏的
destructive 破坏的，毁灭性的，有害的
destructive interference 破坏性干扰，相消干扰
destructive test 破坏性测试

desublimation 去升华作用，凝结（作用）
desulfonation 脱磺酸基
desulphurization 脱硫
desulphurizer 脱硫剂
desultoriness 散漫，不规则
desuperheat 降低过热蒸汽的热量，降温
desuperheater 减温器，过热蒸汽降温器
desurface 除［剥］去……表层，（修整时）清除表层金属
deswell 退［泡，溶］胀
detachability 可拆离性
detached 分离；分离的，分开的
detail sanding 细节砂光，局部打磨
detectability 检测限（敏感度），检测能力
detector 检测系统（检测器），探伤器
detention 延迟，阻止，停滞
deterge 使清洁，去垢，净化
detergence 去垢性，洗净
detergent 清洁剂，净化［洗净，含洗涤剂］的
deteriorate 变坏，降低品质，恶化，损坏，消耗，磨损
deterioration 变质，退化，恶化
determinand 被测定物，待测物
determinate error 可定误差
determination 测定
determine 测定
determining 测定，确定
detersive 清洁剂，去污剂；使清洁的
detorsion 弯曲矫正，曲度不足
detoxication 去除污染
detoxify 去除污染，除害处理
detriment 损害，伤害，不利，造成损害的根源
detrimental 不利的，有害的
detrition 磨损，耗损
detritus 腐质
detritus equipment 破碎设备
detruncate 切去，削去，缩减
detrusion 剪切变形，位［滑］移，压出
detrusion ration 剪切比

developing solvent 展开剂，显影溶剂
deviation 偏差，漂移
devious 偏僻的，弯曲的
devise 设计，发明
devitrification 使不透明，析晶
devitrify 使不透明，除去（玻璃）光泽
devolatilization 脱挥发份作用
dewatering 脱水，浓缩，增稠
dewax 脱［去］蜡
dewetting 去湿，反湿润
dew-pond （人工外成的）存水池
dextranase 葡聚糖酶
dextrinase 糊精酶
dextro- 【构词成分】右旋
dextrorotatory 右旋的，右旋性的
dextrose 右旋糖，葡萄糖
dezincification 脱锌
dezincify 脱［除，去］锌
di- 二（个，重），联（二），双，二倍，分开，（解）除，取消，离，否定
diablastic 筛状变晶（结构）的
diacid 二酸；二酸的，二价酸的
diacolation 渗萃，渗滤
diactinic 透光化线的
diagenism 成岩作用，沉积变质作用
diaglyph 凹雕，凹刻
diagometer 电导计
diagonal 对角线，斜线；斜的，对角线的，斜纹的
diagram 图表，图解；用图解法表示
diagrammatical 图解的，概略的，轮廓的
diagraphy 作图法
dial （刻，标，调谐）度盘，（仪）表面，标度
dialkene 二烯烃
dialkyl 二烃［烷］基的
dialkylene 二烯基
diallyphthalate 苯二甲酸二烯丙酯，邻苯二甲酸二丙烯
dialysis 透析，渗析
dialytic 透［渗］析的，透膜（性）的，分解的，有分离力的
dialyzability 可透［渗］析性
dialyzable 可透［渗］析的
dialyzate 透析液，渗析［出，透］液
dialyze 透［渗］析，渗出，分解（析）
diamagnet 抗磁物质
diamagnetism 逆磁性，抗磁性
diameter 直径
diametral 直径的，（沿直）径（方）向的
diametral compression test 径向受压试验
diametrical 直径的，正好相反的，截然的
diamide 二酰胺，肼
diamido 二氨（酰）基
diamine 二元胺，联氨
diamino- 二氨基
diammine 二氨
diamond cutters 钻石刀具
diamond drilling 金刚石钻孔
diamond grain 金刚砂
diamond polishing 钻石抛光，金刚砂抛光
diamond pyramid hardness 维氏硬度
diamond shot test 金刚石冲击试验法
diamond spar 刚玉，金刚砂
diamond wheel 金刚石磨轮［砂轮］
diamond-cut finish 钻刀处理
diamondite 碳化钨硬质合金
diamondoid 菱形的，钻石形的
diamylene 癸烯，二戊烯
dianion 二阶阴离子
diapause 滞育，间歇期
diaper 菱形花纹［图案］，用菱形花纹装饰
diaphaneity 透明度，透明性
diaphanometer 透明（度）计，色度计
diaphanometry 透明度测定法
diaphanous 透明的，精致的，半透明的
diaphragm 隔膜，横隔膜，隔板
diasolysis 溶胶渗析
diatomaceous earth 硅藻土
diatomic 二原子的，二氢氧基的，二价的
diatomic molecule 双原子分子

diatomite 硅藻土
diazo 重氮化合物；重氮的，二氮化合物的
diazo transfer 重氮基转移
diazoalkane 重氮烷
diazonium coupling 重氮偶联
diazotization 重氮化
diazotization reaction 重氮化反应
diazotization titration 重氮化滴定法
dibutylphthalate 二丁酯邻苯二酸盐
dicarboxylic 双羧酸的，二羧基的
dicarboxylic acid 二羧酸
dichloride 二氯化合物
dichlorinated 二氯化的
dichloroethane 二氯乙烷
dichotomy 二分法，两分，分裂，双歧分枝
dichromate 重铬酸盐
dichromate oxidizability 重铬酸盐需氧量
dichromate value 重铬酸盐值
dichromic 重铬酸的，含两个铬原子的
dichromic acid 重铬酸
dicing 菱形装饰，有菱形装饰之皮革，切割，切成小方块
dicky 易碎的
diclinic 双斜的
dicoelous 双凹的，有两腔的
dicroton 二聚丁烯醛
dicyan （二）氰（基）
dicyanamide 二氰胺
dicyandiamide 双氰胺，氰基胍
dicyanide 二氰化物
dicyanogen 氰（气），乙二腈
dicyclic 二环的，双环的，双周期的
dicyclo- 【构词成分】双环
dicyclopentadiene 二环戊二烯，二茂
dicyclopentadienyl 双茂基
didepside 二缩酚酸
didymous 二重的，双生的
die 模，冲模，钢模
die block 模块，底模

die burn 烧伤（焊接缺陷）
die cast 压铸件
die cast alloy 压铸合金
die casting 压铸，拉模铸造
die casting dies 压铸冲模
die casting machines 压铸机
die cavity 模腔，模槽，阴模
die forging 热模锻，模锻法
die height range 适用模高
die lubricant 模具润滑剂
die pressing 模压
die sinking 刻模
die stamping 浮凸印刷，模压
die structure dwg 模具结构图，模具构造图
dieing out press 冲床
dielectric 电介质，绝缘体；非传导性的
dielectric constant 介电常数，电容率
dielectric displacement 电位移，电介质位移
dielectric insulator 介电绝缘体
dielectric loss 介电损耗
dielectric strength 介电强度
dielectrolysis 渗入电解法
dielectrometry 介电滴定，介电测量（法）
dielectrophore 电介基
dielectrophoresis 介电泳，双向电泳
dieless 无模的
dieless wire drawing 无模拉丝法
diene 双烯，二烯
diepoxides 双环氧化合物
dieresis 分开，切开，离开
dieretic 切开的，分开的，分离的
dies-progressive 连续冲模
diester 二酯，二元酸酯
diethanolamine 二乙醇胺，二个羟基的乙醇胺
diethyl 二乙基的
diethyl ester （二）乙酯
diethyl ether （二）乙醚
diethyl oxalate 草酸（二）乙酯

diethylene glycol 二甘醇
diethylether 二乙醚
differential 微分的，差别的，特异的
differential aeration cell 差异充气电池，氧浓差电池
differential aeration corrosion 氧差腐蚀
differential capacity 微分电容
differential heating 差温加热
differential pulse polarography 示差脉冲极谱法
differential pulse voltammetry 微分脉冲伏安法
differential pulse 差分脉冲，差示脉冲
differential quenching 示差淬火，阶差淬火
differential thermal analysis 示差热分析
differential thermocouple 示差[差分]热电偶
differential thermogravimetry 示差热重量法，微分热重
differentiating effect 均化效应，区分效应
differentiating solvent 区分性溶剂
diffluence 分流，溶解，流出，流动性，溶[融]解，溶[液]化
diffluent 易溶物，潮解；流动性的，易溶解的，溶[液]化的
diffract 衍射，使……分散，碾碎
diffraction 衍射
diffractive 衍射的，绕射的
diffractogram 衍射图
diffractometer 衍射计[仪，器]
diffusant 扩散剂，扩散杂质
diffusate 扩散物，渗出液，弥散物
diffuse 扩[发，播，弥，逸，分]散，传播，漫射；弥漫的，散开的
diffuse double layer 分散双电层
diffuse double layer model 分散双电层模型
diffuse indications 扩散指示
diffuse layer 分散层
diffuse reflection 漫反射
diffusely 广泛地，扩散地
diffuseness 扩散，漫射
diffuser 扩散器，汽化器的雾化装置
diffusibility 扩散性，扩散率
diffusible 扩散性的，可扩散的
diffusion 扩散
diffusion annealing 扩散退火，均匀化退火
diffusion coefficient 扩散系数，扩散率
diffusion constant 扩散常数，弥漫常数
diffusion current 扩散电流
diffusion current constant 扩散电流常数
diffusion flux 扩散通量
diffusion hardening 扩散硬化
diffusion layer 扩散层
diffusion limited current density 极限扩散电流密度
diffusion metallizing 渗金属
diffusion potential 扩散电势，扩散电位
diffusionless transformation 非扩散型相变
diffusiophoresis 扩散电泳
diffusivity 扩散性[率，系数，能力]，扩散率
difluorated 二氟化的
difluoride 二氟化物
dig 挖，掘，卡住，不灵活
digestion 消化
digital 数字的，指状的
digital display 数字显示，数显
digital thermometer 数字式温度计
digital-to-analog converter 数-模转换器
dihalide 二卤化物，二卤化
dihedron 二面体
dihexagonal 复六方的，双六角的
dihexahedron 复六方面体，双六面体
dihydric 二羟基的
dihydride 二氢化物
di-ionic 双离子的
diisoamyl 二异戊基
diisobutylene 二异丁烯
diisocyanate 二异氰酸盐，二异氰酸酯

diisooctyl phthalate 邻苯二甲酸二异辛酯
diketone 双酮，二酮，二元酮
dilapsus 分解，熔解
dilatability 膨胀性
dilatancy 搅胀性，膨胀性
dilatant 膨胀物；膨胀的，膨胀变形的
dilatate 膨胀的；膨胀
dilatation 膨胀，扩张，剪胀
dilatation curve 膨胀曲线
dilate 使扩大，使膨胀，扩大，膨胀
dilation 扩张，扩大，膨胀
diluent 稀释剂；稀释的，冲淡的
dilute 稀释，冲淡
dilute solution 稀溶液
dilute sulfuric acid 稀硫酸
diluted gasket 淡水隔板
diluted 稀释的
dilution 稀释，冲淡
dilution stability 稀释稳定性
dilution test 稀释试验
dimension 尺寸，度量，规格，维
dimension change 尺寸变化
dimensional tolerance 尺寸公差
dimer 二聚体，二聚物
dimeric 二聚的，二部组成的
dimerisation 二聚作用
dimerization 二聚作用
dimethyl 乙烷，二甲基；二甲基的
dimethylbenzene 二甲苯
dimetric 四角形的，四边形的，二聚的
dimi 万分之一
diminish 使减少，减少，缩小，变小
dimorphic 双晶的
dimple 凹坑，表面微凹
dimple fracture 韧窝断口
dimply 有凹处的，凹陷的，有波纹的
ding and dent 钣金凹凸
dinitrate 二硝酸盐
dinitro- 【构词成分】二硝基
dint 凹痕；击出凹痕
dioctylphthalate 钛酸二辛酯

diol 二醇
diolefinic 二烯的
diosmosis 渗透，交渗作用
dioxide 二氧化物
dip 下沉，下降，倾斜，浸渍，蘸湿；浸，下降，下沉，泡，蘸
dip coating 浸涂，浸渍涂覆
dip mold 浸渍成形，沉模
diphase 双相的，复相的
diphasic 二相的，双相的
diphenyl 联苯，二苯基
diphosphate 二磷酸盐（酯），磷酸氢盐
dipl- 【构词成分】二重，双，复
diplo- 【构词成分】二重，双，复
dipolar 两极的
dipolar coordinates 双极坐标
dipolar ion 偶极离子
dipolar polarizability 偶极子极化率
dipolarity 偶极性
dipole 偶极子
dipole moments 偶极矩
dipole orientation polarization 偶极子取向极化
dipolymer 二元共聚物，二聚物
dipping 浸渍
dipping tank 浸槽
diradical 双自由基，双游离基
direct current anodizing 直流阳极氧化法
direct current plasma emission spectrometer (DCP) 直流等离子体发射光谱仪
direct hot air oven 直接加热式热风烘干炉
direct potentiometry 直接电位法
direct printing 印刷染色法
direct quench aging 直接淬火时效
direct redrawing 顺向再拉延
direct reduction process 直接还原法［过程］
direct stress 正交应力，法向应力
direct-arc furnace 直热电弧炉
directional property 方向性
dirt 颗粒，污垢

dis- 【构词成分】不，相反，相对，分离，夺去
disable 使失去能力，禁止使用，使无用
disacidify 去酸，脱酸，将酸中和
disaggregate 分解，使……崩溃，解开聚集，解聚；无组织的，分解的
disaggregation 解聚作用，解聚
disalignment 偏离轴心，未对准
disassemble 拆开，解开
disazo 二重氮，重氮基
discal 盘状的，圆盘的
discale 碎鳞，除鳞
discerp 扯碎，撕裂，分开
disc sander 盘式打磨机
discharge 放电，排放量，排放
discharge capacity 放电容量，排流能力
discharge characteristics 放电特性
discharge current 放电电流
discharge current density 放电电流密度
discharge curve 放电曲线
discharge depth 放电深度
discharge duration time 放电持续时间
discharge in high rate current 标称电压，大电流放电
discharge voltage 放电电压
discharge watt-hour 放电瓦时
discharged ampere-hour 放电安时
disco- 【构词成分】盘状，盘
discoloration 变色，掉色
discontinuous eutectic 不连续共晶
discontinuous precipitaion 不连续析出
discrete 离散的，不连续的
discrete sampling 不连续采样，离散抽样
disembogue 使倾注，流入，倾注，流出
disgorge 吐（出），放［喷］出，放弃，除去沉淀物
disinfectant 消毒剂；消毒的
disintegration 瓦解，崩溃；裂变，蜕变
disintegrator 粉碎机
dislocation 位错
dislocation density 位错密度
dislocation free crystal 无位错结晶
dislocation line 位错线
dislodge 逐出，驱逐，使……移动，用力移动，离开原位
dislodge sludge 沉积泥渣
dislodger 沉积槽
dismantle 拆除，取消，解散，除掉……的覆盖物，可拆卸
dismetria 不对称运动
dismutation 歧化作用
disnature 使……不自然，使……失去自然形态
disodic 二钠的
disodium 二钠的，分子中有两个钠原子的
disomatic 二晶质的，捕获晶的
disoperation 侵害作用，相害作用
disordered solid solution 无规律固溶体，无序固溶体
disorder 无规律，无序，失常，缺陷
disordering 无序化
disordering effect 无序化效应
disorganization 混乱，无组织，结构破坏
disoxidation 脱氧，还原，减氧作用
disparity 不同，不一致，不等
dispart 使分开，使分离，分离，分开
dispel 驱散，驱逐
dispensation 分配，配方，处理，免除
dispense 分配，分发，免除，执行
dispenser 分配器，药剂师，施予者，分配者
dispenser cathode 储备式阴极
dispergation 解胶，胶液化作用
dispergator 解胶剂，胶溶剂
dispersancy 分散力，分散性
dispersant 分散剂，扩散剂
dispersate 分散质
disperse 分散，使散开，传播；分散的
dispersed 弥散的，散布的，被分散的，被驱散的
dispersed medium 分散介质，弥散剂
dispersed part 弥散相，分散质

dispersed phase 弥散相
dispersed phase hardening 弥散硬化
dispersed substance 分散物质，分散质
dispersion 分散
dispersing power 分散力，分散能力
dispersion medium 分散介质
dispersion strengthening 弥散强化，第二相强化
dispersion system 分散系
dispersion toughening 弥散增韧
dispersity 分散性，分散度
dispersive 分散的，弥散的
dispersoid 弥散体，分散胶体
displaceable 可替换的，可移置的，可取代的
displaced 位移的，被取代的
displacement 位移，错位，移位
displacement energy 位移能
displacement spike 原子移位区，位移尖峰
displacer 置换剂，取代剂
disposable 可任意处理的，可自由使用的，用完即可丢弃的
disposable load 自由载量，可卸载荷
disposal 处理，支配，清理，安排
dispose 处置，处理，安排
disposed goods 处理品
disposition 处置，配置，配备，布置，安排
disproportion 不均衡，不相称；使不均衡
disproportional 不相称的
disproportionate 不成比例，歧化，不相称的
disproportionation 歧化反应，不均匀
disputable 有争议的
disqualification 不合格，不适合，无资格，取消资格
disregistry 错合度，错配度
disrupt 破坏，使中断；分裂的，中断的，分散的
disrupted horizon 变位层，断错层位
disruptive 破坏的，分裂性的

disruptive discharge 击穿放电，火花放电
disruptive strength 破裂强度，击穿强度，介电强度
disruptive test 击穿实验
disruptive voltage 崩溃电压，击穿电压
disruptiveness 破裂性，分裂
disrupture 破裂，分裂，毁坏
dissect 切细，仔细分析；进行解剖，进行详细分析
dissected 切开的，分开的，多裂的
dissimilar 不同的，相异的，不相似的
dissimilar metal corrosion 异极金属腐蚀
dissipate 驱散，使消散，消散
dissipation factor 损耗系数，耗散因数，损耗因子，功耗因素
dissipative 浪费的，消耗的，消散的
dissipative element 耗散元件
dissociable 可以离解的，可分离的，易解离的
dissociate 离解
dissociated 离解的，游离的，分裂的
dissociation 离解
dissociation constant 解离常数，电离常数
dissociation pressure 解离压力
dissociative 游离的，分离的
dissociative electron transfer 离解电子转移
dissolubility 溶解度，可溶性，溶解性
dissoluble 可分解的，可溶解的
dissoluent 溶剂
dissolution 分解，溶解
dissolvability 溶解度，可溶性
dissolvable 可溶解的，可分解的，可解散的
dissolvant 溶剂，溶解，溶媒
dissolve 使溶解，使分解，使液化，溶解
dissolved 溶解的，溶化的
dissolved impurity 溶解杂质
dissolved oxygen 溶解氧
dissolvent 溶剂，有溶解力的
dissolver 溶解装置，溶解器

dissolving metal reduction 溶解金属还原
dissolving power 分解能力，分辨能力
dissymmetrical 非对称的，不对称的
dissymmetry 不对称，反对称
distal 末梢的，末端的
distance 距离
distemper 水浆涂料，色胶，胶画（颜料），刷墙粉
distend 使……膨胀，使……扩张，扩张，膨胀
distensibility 膨胀性，扩张性
distensibility meter 杯突试验机
distensible 扩大的，可扩张的，胀的
distent 扩张；膨胀的
distention 膨胀，扩张
distichous 二列的，二分的，分成两部分的
distill 馏出物，馏分；蒸馏，渗出
distillate 蒸馏物，馏分
distillation 蒸馏，净化，蒸馏法，蒸馏物
distillatory 蒸馏用的
distilled water 蒸馏水
distilled 蒸馏的
distilling 蒸馏的，蒸馏作用
distinct 明显的，独特的，清楚的，有区别的
distinctness of image 平滑度及鲜映性指标
distort 扭曲，使失真，曲解
distortion 变形，歪变，畸变
distortional 变形的，歪曲的，畸变的
distortionless 无畸变的，不失真的，无失真
distributing 分配的
distributing groove 配水槽，布水槽
distributing hole 配水孔，聚水孔
distribution 分布，分配
distribution coefficient 分配系数，分布系数
distribution of molecular weight 分子量的多分散性
distribution ratio 分配比

distributive 分配的，分布的，分发的
distributivity 分配性，分布性，分配律
distributor 分配器，配电盘
district 区域，地方；把……划分成区
district heating 局部加热
distrub 扰乱，干扰，使紊动，打乱，妨碍
distrub current cycle 干扰电流周期
distrub output 干扰输出
distrub area 受扰区
distrubed sample 扰动样品，非原状样品
distrubing current 干扰电流
distrubing effect 干扰效应，紊乱效应
distrubing force 扰动力
distrubance 扰动，干扰，故障，障碍，失调，破坏，妨碍，损伤，变位
distrubed 干扰的，扰动的
disubstituted 双取代的，二基取代的
disubstitution 双取代作用
disulphate 硫酸氢盐，酸式硫酸盐，焦硫酸盐
disulphid 二硫化物
disulphonate 二磺酸盐
disulphonic acid 二磺酸
disulphuric acid 焦硫酸
disymmetrical 双对称的
disymmetry 双对称，对映形态
dit 小孔，砂眼
ditch 沟渠，壕沟；在……上掘沟，把……开入沟里，丢弃，开沟，掘沟
ditchwater 沟水
ditetragon 双四边形
ditetragonal 复正方形的
ditetrahedron 双四面体
dither 高频脉动，颤动调谐
dithiane 二噻烷
dithiocarbonate 二硫代碳酸盐
dithionate 连二硫酸盐
dithionite 连二亚硫酸盐
dititanate 二钛酸盐
ditolyl 二甲苯基，联甲苯

ditrigon 双三角形
ditrigonal 复三方的
divacancies 双空位，双空格点
divalence 二价
divalent 二价的
divariant 二变的，双变的
diverge 使偏离，使分叉；分歧，偏离，分叉，发散，逸出，散射
divergence 分歧，发散，逸出，散射，扩张
divergence angle 发散角，扩张角
divergence coefficient 发散系数，漏损系数
divergent 相异的，分歧的，散开的，辐射状的，非周期变化的，扩张的
divergent nozzle 扩散喷嘴
divergent series 发散级数
diverging 发散的，分歧的，岔开的
diverging faults 枝状断层
diverging lens 发散透镜
divers 不同种类的，各式各样的
diverse 不同的，多种多样的，变化多的
diversification 多样化，变化
diversion 转换，转向，变向，导流，分出，引出
diversity 多样性，差异
divert 转移，使……转向
divide 划分，除，分开
dividual 分开的，可分开的，分配的，独特的
divinyl 二乙烯基，联乙烯，丁二烯
division 分割，划分，分度，隔板，阻挡层，间隔
dipotassium tartrate 酒石酸二钾
dodec (a) - 【构词成分】十二
dodecagon 十二边形
dodecahedral 十二面体的
dodecahedron 十二面体
dodecane 十二烷
dodecanol 十二烷（醇）
dodecene 十二烯

dolicho- 【构词成分】长
dolichomorphic 长形的，狭长的
domain 磁畴，界，圈
domatic 坡面的
dome 凸圆，成圆顶状
domelike 穹顶（状）的
domical 圆顶（式）的，穹隆式的
domy 圆顶的
donor 供［授，给］体，施主（杂质）
donor level 施主能级
dopant 掺杂剂，掺杂物
dope （机翼的，航空用，涂布）漆，涂料，涂布油，明胶
dope chemical 掺杂剂，掺杂元素
doped coating 加固涂料，涂漆包皮
doped glass 掺杂玻璃
doping 掺杂质，加添加剂，涂上航空涂料
doping defect 掺杂缺陷
doping profile 掺杂分布，掺杂剖面
doping system 掺杂系统
dose 剂量
dosing ratio 投配比，投配率
double action sander 双动打磨机
double amperometric titration 双电流滴定法，双安培滴定法
double annealing 双重退火，两步退火
double bond migration 双键移位
double contraction 双重收缩
double current pulse 双电流脉冲法
double equilibrium state 复平衡状态
double indicator titration 双指示剂滴定法
double layer nickel coating 双层镍（涂层）
double potential step chronamperomery 双电势阶跃计时电流（安培）法
double potential step chronocoulometry 双电势阶跃计时库仑法，双电势阶跃计时电量法
double quenching 双重［两次］淬火
double salt 复盐
double scattering 双散射，二次散射
double side lapping machine 双面磨光机

double side laser engraving 双面激光雕刻
double side polishing machine 双面抛光机
down stream plasma etching system 分离型等离子体蚀刻系统
downward 向下；向下的，下降的
dragout 废酸洗液
drain 排水管，排水沟，排水道，排水
drainage 排水，排水系统，污水，排水面积
drape 打折（覆盖）
draught 气流
dark color 深色
draughty 通风的，通风良好的
drawability 可拉性，回火性，压延性
drawback 缺点，不利条件
drawing 图样，拔制，拔出
drawing deep 深拉延，深度引长
drawing die 拉模，拉延模，深冲模
drawing forming 压延成型
drawing metals 拉拔金属
drawing polymers 牵伸聚合物
drawing strenghtening 拉拔强化
drawing wet 湿拉延
drawing compounds 拉丝乳剂
dredger 挖泥船，疏浚机，挖泥机，撒粉器
dregginess 污浊的，有渣滓的
dreggy 污浊的，有渣滓的，多渣的，混污（浊）的
drench 浸液，使湿透
dresser 砂轮整修机，砂轮修整器
dressing 修整，表面修光
dribbing 零星修补
dribble 使滴下，滴流，细流，
drier stabilizer 催干稳定剂
drier 干燥剂，催干剂
drift 漂移
drill 钻
drill stand 钻台
drilling machine 钻床
drilling machine bench 钻床工作台

dripping 流痕
driving force 驱动力，推动力
drop feed carburizing 滴注渗碳
drop forging 冲锻法，锤锻法，模锻法
drop test machine 跌落试验机
dropping corrosion test 点滴腐蚀试验
dropping mercury electrode（DME） 滴汞电极
dross 渣滓，浮渣，碎屑
drossy 铁渣的，碎屑的，含有浮渣的
drown 淹没，浸没，浸湿，使湿透
drum sander 滚筒砂光机
dry contamination 干污染
dry etching system 干式蚀刻系统
dry sandpaper 干打磨用砂纸
dry spray booth 干式喷漆室
dry through 完全干燥，完全固化
dryer 烘干机，干燥剂
drying 干燥
drying agent 干燥剂
drying equipment 干燥设备
drying by evaporation 蒸发干燥，挥发干燥
drying off 烘干水分
drying oil 干性油
drying oven 干燥室，干燥箱
ducking 浸入水中，湿透
duct 风筒，输送管，导管
ductibility 延展性，可锻性，可塑性
ductible 可延展的，可锻的，可塑的
ductile 延展性，延性的，易变性，柔软的
ductile cast iron 球墨铸铁
ductile crack 延性破裂
ductile fracture 塑性破坏，塑性断口，韧性断裂，塑性断裂，延性破裂
ductile iron 球墨铸铁，延性铁，韧性铁
ductile rupture 延性破坏，韧性断裂
ductile-to-brittle transition 延性-脆性转变
ductility 延展性，延度，可锻性
ductility test 延展性测试
ductility transition 延性转变，韧性转变

dull 缓和，减轻，迟钝的，无光泽的
dull deposit 毛面镀层
dull grey 哑灰
dulling 失光，发暗，迟钝
dumped 废弃的，弃扔的，堆积的
dumped packing 乱堆填料，散装填料，堆积填充物
dunting 冷裂，风裂
Duolite 离子交换树脂
duoplasmatron 双等离子管，双等离子体发射器
duplet 电子对，粒子对
duplex 二倍的，双重的
duplex process 双重熔解法
duplex pulse 双脉冲
duplicate 副本，复制品；复制的，二重的；复制，使加倍
duplication 副本，成倍，复制
duplication check 双重校验，重复校验
duplicative 加倍的
duplicity 二重性，互换性
durability 耐久性，持久性
durability test 耐久性试验，疲劳试验
durable 耐用的，持久的
duration 持续（时间），延续（性，时间），耐久，耐用
durative 持续性；持续的，持续性的
dust 灰尘，尘埃［土］
dust coat 粉涂
dust collector 集尘器
dust gauze 滤灰网，过滤网
dust gun 手摇喷粉枪，手摇喷粉器，手持喷雾器
dust keeper 防尘装置，集尘器，防尘板
dust proof 防尘
dustlike 尘状的，粉状的
dusty 积满灰尘的，粉末状的，灰尘的
duty-cycle 工作循环的
dvi- 【构词成分】类，似，第二的
dwelling 停止，保压
dwelling period 停止周期
dwindle 减少，变小，使缩小，使减少
dyad 双，一对，二价元素；二的，双的
dye 染料，染色；染，把……染上颜色
dye check 着色检查，着色探伤
dye gigger 精染机
dye penetrant inspection 着色渗透检验，着色探伤，着色检验，染料渗透检查
dye spot test 染斑试验
dye strength 染料浓度
dyeability 可染性，可着色性
dyeable 可染色的
dyeing 染色，染色工艺
dyeing ability 上色性
dyeing power 染着力，着色能力
dyestuff 染料，着色剂
dynamic 动态，动力；动态的，动力的，动力学的，有活力的
dynamic emittance matching 动态发射匹配
dynamic hardening 动态硬化
dynamic imitation test 动态模拟试验
dynamic modulus of elasticity 动弹性模数
dynamic ohmic drop compensation 动态欧姆降校正
dynamic range 动态范围
dys- 【构词成分】恶化，不良，困难，障碍

E

early transient regime 暂态区，早期不稳定阶段
earthing device 接地装置
earthing pole 接地极
earthy element 土族元素
earthy water 硬水
earthy 土的，土质的，泥土的，接地的
easily damaged part 易损件
eat-back 回蚀，(化学腐蚀)蔓延
ebb 衰退，衰落；减少
ebonite 硬橡胶，硬化橡皮，硬橡皮
ebullator 沸腾器，循环泵
ebullience 沸腾，冒泡，起泡，充溢，爆发
ebulliometry 沸点测定（法）
ebullioscopic 沸点测定器的
ebullioscopy 沸点升高测定法
ebullition 沸腾，冒泡
eccentricity 离心率，偏心度，偏心距
eccentric 偏心盘；偏心的，离心的
ecclasis 脱落，破碎
eccysis 洗除
echelon 梯形，梯次编队，梯阵，阶层；排成梯队
echo depth sounding sonar 回声测深声呐，回声测深仪，回波声呐
echo elimination 回波消除
echo 回波，反射，重复
echolocation 回波定位，回声测距，回声定位能力
echosonogram 超声回波图
ecocycle 生态循环
ecodeme 生态同类群
ecological distribution 生态分布
ecological equilibrium 生态平衡，生态均衡
economical 经济的，节约的，合算的
economizer bank 预热管，节热器排管

economizer 节约装置
ecouvillon 硬刷子
ecphlysis 破裂，绽裂
ecphylactic 无防御的
ecphylaxis 无防御性
ecstaltic 离心的
ectad 向外，在外面
ectal 外侧的，表面的，外表的
ectasia 扩张，膨胀
ectasy 扩张，膨胀
ectatic 扩张的，膨胀的
eddy current flaw detector 涡流探伤仪
eddy current gages 涡电流测厚计
eddy current testing 涡流检测，涡流探伤
eddy current 涡流
eddy defect detector 涡流探伤机
eddying 涡流；涡流的，旋转的
eddy 涡流，漩涡，逆流；旋转，起漩涡
edge analysis 图像边缘分析
edge bending 侧向弯曲
edge crack 边缘裂纹，裂边
edge dislocation 刃型位错，边缘位错
edge echo 棱边回波
edge effect 边缘效应
edge polisher 边缘抛光机
edge wave 边缘波，边波
edge 边缘，棱边
eduction 引［抽，排，析，提，流］出，离析；引出物
educt 离析物，游离物
edulcorate 洗净，使纯，纯化，从……除去酸类［盐类、可溶性物质］
effect 影响，效果，作用；产生，达到目的
effect lacquer 美饰漆
effective 有效的，起作用的，实际的，实在的
effective case depth 有效硬化深度

effective depth penetration 有效穿透深度，有效透入深度
effective hardening depth 有效淬硬深度
effective magnetic permeability 有效磁导率
effective nuclear charge 有效核电荷
effective permeability 有效磁导率，有效渗透率，相对渗透率
effective reflection surface of flaw 缺陷有效反射面
effective resistance 有效电阻
effective strain 有效应变
effective stress 有效应力
efferent 传出的，输出的
effervescent 冒泡的，沸腾的
effervesce 泡腾，冒泡
effervescing steel 沸腾钢，净缘钢
efficacious 有效的
efficacy 功效，效力
efficiency 效率，效能，功效
efficient 有效率的，有能力的，生效的
efflorescence 粉［风，晶］化，吐霜，泛白
efflorescent 风化的，起霜（指盐霜）
effloresce 风化
effluence 流出，流出物，发射物
effluent 出水，流出物，污水，工业废水；流出的
effluent control 废水及废气控制，流出物控制
effluent disposal 废水处理
effluent fraction 流出的馏分
effluent gas 废气，烟道气
effraction 破裂，裂开，弱化
effumability 易挥发性，挥发率
effumable 易挥发的
effuse 流［涌，泻，喷，渗，泄］出
effusive 流［涌，泻，喷，渗，泄］出的
eigen- 【构词成分】本征，特征，固有
ejection 喷出，排出物

ejection nozzle 喷嘴
elastic 有弹性的，灵活的，易伸缩的
elastic constant 弹性常数
elastic deformation 弹性形变
elastic fatigue 弹性疲劳
elastic hardness 弹性硬度
elastic hysteresis 弹性迟滞
elastic medium 弹性介质
elastic modulus 弹性模数，弹性系数
elastic property 弹性性质，弹性特性
elastic recovery 弹性回复，弹性复原
elastic scattering 弹性散射
elastic strain 弹性应变，弹性形变
elasticizer 增塑剂，弹性增进剂
elastomer 高弹性聚合物，弹性体
elcometer 永久磁性测厚仪，膜厚测定仪
electric 电的，电动的，发电的，导电的
electric arc furnace 电弧炉
electric coloring zinc 电彩锌
electric conductivity 导电性，导电率
electric conductor 导（电）体
electric density 电荷密度
electric dipole 电偶极子
electric double layer 双电层
electric etcher （金属）电解腐蚀器
electric field 电场
electric forming 电成型，电冶，电铸
electric furnace 电炉
electric immersion heater 浸没式电热器
electric induction 电感应
electric inductivity 电感应性，介电常数
electric resistance furnace 电阻炉
electric resistance heater 电阻加热器
electric resistance thermometer 电阻温度计
electric valve 电动阀，电动阀门，电磁阀，整流管
electric wave filter 滤波器
electrical conductance gag 电传导测厚计
electrical conductivity 导电性，导电率

electrical dipole 电偶极
electrical discharge machining (EDM) 放电加工，火花电蚀法
electrical material 电气材料
electrical property tester 电性能测定仪
electrical resistivity 电阻率
electricity metallurgy 电冶金
electrification 电气化，带电，充电
electrify 使电气化，使充电，使触电
electrion 高压放电
electrization 电气化，带电法
electro 电镀物品；电镀
electro capillary curve 电毛细曲线
electro corrosion 电化腐蚀
electro extraction 电解萃取，电解提纯
electro migration 电迁移
electro neutrality 电中性
electro phoresis 电泳
electro plating 电镀
electro-adsorption 电吸附
electro-equivalent 电化当量
electro-etching 电解侵蚀，电蚀刻法
electro-hydrometallurgy 电湿法冶金
electro-ultrafiltration 电超滤，电渗析
electroaffinity 电亲和势，电亲和力，电亲和性
electroanalysis 电解，电分析
electrobath 电镀浴，电解槽，电镀用液
electrobrightening 电解光亮化
electrocapillary curve 电毛细管曲线
electrocapillary equation 电毛细方程
electrocapillary measurement 电毛细曲线测量法
electrocatalysis 电催化作用
electrochemical 电化学的
electrochemical analysis 电化学分析
electrochemical cell 电化学电池
electrochemical corrosion 电化学腐蚀
electrochemical equivalent 电化学当量
electrochemical machining 电化学加工
electrochemical methods 电化学分析法

electrochemical noise analysis 电化学噪声分析
electrochemical noise (EN) 电化学噪声
electrochemical overpotential 电化学过电位
electrochemical oxidation 电化学氧化
electrochemical polarization 电化学极化
electrochemical potential 电化学势，电化电位
electrochemical reduction 电化学还原
electrochemical spectroscopy 电化学谱
electrochemical work station 电化学工作站
electrochemically modulated infrared spectroscopy 电势调制红外反射光谱学
electrochemiluminescence 电化学发光
electrochemistry 电化学
electrocoating equipment 电泳涂装装置
electroconductive additives 导电剂
electrode 电极
electrode frame 极框
electrode potential 电极电位，电极电势
electrode reaction 电极反应
electrodecantation 电倾析
electrodecomposition 电解分解作用
electrodeionization 电极电离作用，去电离子
electrodeless 无电极的
electrodeposit 电沉淀物，电沉积
electrodeposition coating 电沉积膜
electrodeposition 电沉积，电镀
electrodesiccation 电干燥法
electrodialysis 电渗析
electrodiffusion 电扩散（系数）
electrodispersion 电分散，电分散作用
electrodissolution 电溶解，电解溶解
electrodissolvent 电解溶解剂
electroduster 静电喷粉器
electrodusting 静电喷粉
electrodynamical 电动力的，电动力学的
electroendosmosis 电内渗，电渗透，电渗（现象）

electroengraving 电刻术（物）
electroerosion 电腐蚀，电侵蚀
electroexcitation 电致激发
electroextraction 电解提取［萃取，抽出］
electrofiltration 电滤，电滤作用
electrofission 电致裂变
electrofluorescence 电致发光
electroforming 电铸
electroform 电铸［冶，成型，沉积］
electrogalvanize 用锌电镀；在……上电镀锌
electrogenerated chemiluminescence 电致化学发光
electrography 电蚀刻，电刻术，电谱法
electrogravimetric analysis 电重量分析法，电重量法
electrogravimetry 电重量分析法
electrograving 电刻，电蚀刻
electrokinetic phenomena 界面电动学现象，电动现象，动电现象
electrokinetic 动电的，动电学的
electroless nickel 化学镀镍
electroless plating 化学镀，非电解镀层
electroless 无电镀的，无电的，化学镀
electroluminescence 电致发光
electrolysate 电解产物
electrolysis 电解，电解作用，电蚀
electrolyte 电解液，电解质，电解
electrolyte hydrometer 电解液比重计
electrolyte level control pipe 电解液液面控制管
electrolyte level indicator 电解液液面指示器
electrolyte level sensor 电解液液面传感器
electrolytic 电解的，电解质的，由电解产生的
electrolytic analysis method 电解法
electrolytic capacitor 电解电容器，电解电容，电解质电容器
electrolytic cell 电解池
electrolytic charger 电解充电器，电解液充电器
electrolytic cleaning 电解清洗
electrolytic colored anodic oxide coating 电解染色阳极氧化膜
electrolytic colouring (EC) 电解着色
electrolytic corrosion 电解腐蚀，电解侵蚀，电蚀
electrolytic current 电解电流
electrolytic degreasing 电解除油
electrolytic deposition 电解沉积，电镀
electrolytic dissociation 电离，电解离解，电离解
electrolytic etching 电解蚀刻
electrolytic grinding 电解研磨
electrolytic hardening 电解淬火
electrolytic ion 电解离子
electrolytic pickling 电解侵蚀
electrolytic polishing 电解抛光
electrolytic refining 电解精炼
electrolytic sliver recovery unit 电解银回收装置
electrolytic solution 电解液
electrolytic stream etch process 电解流腐蚀过程
electrolytic titration 电位滴定法
electrolytics 电解学，电解化学
electrolyzable 可电解的，易电解的
electrolyzation 电解［离］
electrolyze 电解，用电分解，使电解［离］
electrolyzer 电解剂，电解器［池，装置，槽］
electromachining 电加工，电火花加工
electromagnet 电磁铁，电磁体
electromagnetic acoustic transducer 电磁超声换能器
electromagnetic induction 电磁感应，电磁效应，感应电流
electromagnetic radiation 电磁辐射
electromagnetic spectrum 电磁波谱
electromagnetic testing 电磁检测

electromagnetic 电磁的
electromatic 电气自动的，电气自动方式，电控自动方式
electromechanical polishing 电解机械抛光
electromechanical 电动机械的，机电的
electrometallurgical 电冶金的
electrometer 静电计，静电测量器，电位计，量电表
electrometric titration 电滴定，电势滴定电，位滴定法
electrometric 测电术的，电测量的，测电的
electrometry （静电计）测电术，电测法（术），量电法
electromicrograph 电子显微照相
electromicroscope 电子显微镜
electromigration 电迁移
electromigratory 电迁移的
electromobility 电迁移率
electromotive force（EMF） 电池电动势
electromotive series 电动序，元素电化序
electromotive 电动的，电动势的
electromotor 电动机，发动机，发电机
electron 电子
electron affinities 亲电性，电子亲和能
electron attracting group 吸电子基
electron beam curing 电子束固化，辐射固化漆膜，辐射交联
electron capture detector 电子捕获检测器，电子俘获探测器
electron configuration 电子构型，电子组态，电子排布
electron configurations in octahedral complex 八面体构型配合物的电子分布［构型］
electron diffraction 电子衍射
electron donating group 供电子取代基
electron with-drawing group 吸电子取代基
electron energy band 电子能带
electron energy disperse spectroscopy 电子能谱仪
electron energy loss spectroscopy（EELS） 电子能量损失谱
electron impact source 电子轰击离子源
electron lens 电子透镜
electron micrograph 电子显微照片，电子显微放大器
electron microscope 电子显微镜
electron paramagnetic resonance（EPR） 电子顺磁共振，电子自旋共振
electron paramagnetic resonance 电子顺磁共振
electron photomicrograph 电子显微照片
electron probe microanalysis（EPMA） 电子探针显微［微量］分析
electron probe microanalyzer 电子探测显微分析仪，微量分析器，电子探针微量分析器
electron ratio 电子比
electron radiography 电子辐射照，电子辐射照相术
electron shading effect 电子遮掩效应
electron spin resonance 电子自旋共振，顺磁共振
electron spin 电子自旋
electron state 电子态
electron volt 电子伏特
electronate 使增电子，使还原
electronating agent 增电子剂，还原剂
electronation 增电子作用，还原作用
electronegative 带负电的，负电性的，亲电子的
electronegativity 电负性
electroneutrality 电中性
electronic conductor 电子导体
electronic noise 电子噪声
electronic polarization 电子极化
electronic transmission coefficient 电子传输［传递］系数
electronic 电子的
electroosmosis 电渗，电渗透，电渗析
electropherography 电色谱法，载体电泳

图法
electrophile 亲电体，亲电子试剂
electrophilic addition 亲电加成
electrophilic reagent 亲电试剂
electrophilic rearrangement 亲电重排
electrophoresis 电泳
electrophoresis coating 电泳漆，电泳涂料
electrophoretic 电泳的
electrophoretic coating 电泳膜，电泳涂敷，电泳喷涂
electrophoretic deposition 电泳沉积
electrophoretic force 电泳力，电渗力
electrophoretic powder coating 粉末电泳涂装法
electrophoretogram 电泳图
electrophotometer 光电比色计，光电光度计，光电计
electrophotophoresis 光电泳
electroplating 电镀
electropolishing 电解抛光
electropositive 正电的，阳性的
electropositive potential 正电性电位
electrostatic air atomizing spray method 空气雾化式静电涂装
electrostatic coating 静电涂装，静电喷涂层
electrostatic coating equipment 静电涂装装置
electrostatic discharge protection 静电放电保护
electrostatic fluidized bed 静电浮床，静电流化床
electrostatic potential 静电势，静电位
electrostatic powder coating equipment 静电粉末涂装装置
electrostatic precipitation 静电沉积
electrostatic scan 静电扫描
electrostatic separation 静电分离
electrostatic spraying 静电喷涂，静电喷射
electrostatic 静电的，静电学的，静电式
electrothermal 电热的

electrothermal equivalent 电热当量
electrothermic method 电热法
electrothermic 电热的，电致热的
electrotitration 电滴定
electrotype 电铸板
electrotyping 电成型，电铸术，电铸，电铸法
electrovalent compound 电价化合物
element 化学元素，成分
elementary reaction 基元反应，单元反应
elementide 原子团
eleo- 【构词成分】油
eleometer 油度计，油比重计
eletromotive force 电池电动势
elevation 正视图，立视图
eliminate 消除，排除
elimination reaction 消除反应
eliquation 熔化［析，解］，液析，偏析
elixiviation 浸滤，浸析，去碱
ellipsometry 椭圆偏振法，椭圆光度法，椭圆偏光法
elliptically 呈椭圆形
elongation 延展性，伸长度
eluant 洗脱剂，洗提液，洗脱液
eluate 洗出液，洗脱物，洗脱液
eluent 洗脱液，洗提液
elute 洗提
elution 洗脱，洗提
elutriant 洗提液
elutriation 淘析
eluvial 渗蚀层，残积的，残积层的，淋滤的
eluviate 淋溶
eluviation 淋溶作用，残积作用
embedded 植入的，深入的，内含的；把……嵌入，埋入
embedded part 预埋件，嵌入
emboss 装饰，使凸出，在……上作浮雕图案
embossing 浮花压制加工，压花
embossment 浮雕，凸起

embowed 弓形的，弧形的
embrittle 使变脆
embrittlement 脆化
embrocation 擦剂，涂擦剂
emery 金刚砂，刚玉砂
emery cloth 金刚砂布
emery paper 金刚砂纸
emergizer 促进剂（渗碳）
emission （光、热等的）发射，散发，喷射
emission electron microscope 发射电子显微镜
emissivity 发射率，辐射系数
emittance 发射率，发射量，辐射强度
emitter 发射极，发射器
emollescence 软化，熔融
emolliate 软化
emollient 软化剂，润滑剂；使柔软的
emphasis 重点，强调
emphasis circuit 校正[补偿]电路
empaistic 浮雕的，（装饰品）有隆地花纹的
empire cloth 绝缘油布，漆（胶）布；黏性物质
emplastic 黏性的，黏合的，胶合的
employ 采用
empoison 污染，使……中毒
emulphor 乳化剂
emulsibility 乳化性，乳化度
emulsible 可成乳状的，能乳化的
emulsicool 乳浊状油（切削液），乳化液，
emulsifiable 可成乳状的，能乳化的
emulsification 乳化（作用）
emulsified oil quenching 乳化油淬火
emulsifier 乳化剂
emulsify 使……乳化，乳化
emulsifying agent 乳化剂
emulsifying liquid 乳化液体
emulsion 乳剂，乳状液
emulsion cleaner 乳化清洁剂
emulsion degreasing 乳化除油，乳化脱脂法

emulsion liquid membrane 乳化液膜
enamel 搪瓷，瓷漆；彩饰，涂以瓷釉
enamine 烯胺
enantiomorph 对映体，对映结构体
encapsulation 封装
encapsulation molding 注入成形，低压封装成型
encircling coils 环形线圈，环绕式线圈
encircling diffusion 环状扩散
enclitic 附属的，斜面的
enclosed 被附上的，封闭式，封闭的
enclosure 夹杂物，附件，外[包]壳
encrimson 红色涂料[油漆]；使成深红色
encroach 侵占，侵蚀，侵犯
encroachment 侵入[犯，蚀，害]
encrust 包上外壳，装饰外层，结壳
encumbrance 妨碍[害]，阻碍（物），障碍物
end absorbtion 末端吸收
end elevation 侧视图，端视图
end effect 端部效应 端点效应
end instrument 终端设备，传感器，敏感元件
end quenching test 端淬试验
end socket 端头，（钢索的）封头
endless grinding belt 循环式研磨带
end-of-charge voltage 充电结束电压，充电终止电压
end-off voltage 放电截止电压
endoenergic 吸能的
endoergic 吸能的，吸热的
endophilicity 亲和性
endosmic 渗入的，内渗的
endosmose 内渗，电内渗，渗透
endosmotic 内渗（性，现象）的
endotaxy 内延，内整向
endotherm 吸热
endothermal 吸热[能]的，吸热反应的，内热的
endothermic 吸热的，吸能的
endothermic and exothermic processes 吸热

与发热过程
endothermic atmosphere 吸热式气氛
endothermic reaction 吸热反应
endurance 耐力，持续时间，忍耐，持久性，持续性
endurance crack 疲劳断裂
endurance expectation 估计使用期限
endurance failure 疲劳破坏，疲劳破裂
endurance limit 疲劳极限，持久限度
endurance strength 疲劳强度
endurance test 疲劳试验
enduring 持久的，能忍受的，耐磨的，耐用的
Enduro 铬锰镍硅合金，镍铬系耐蚀耐热钢
enediol 烯二醇
energetic-ion analysis（EIA） 荷能离子分析
energetics 热力学，动能学，力能学
energies of orbital 轨道能量
energization 通电，激发
energize 激励，使活跃，供给……能量，活动，用力
energizer 激发器，增能器，渗碳加速剂，（尿素树脂）硬化剂，抗抑制剂
energy band 能带
energy disperse spectroscopy（EDS） 能谱仪
energy gap 能隙
energy packet 能，束
energy recovery 能源再生
enfoldment 折叠
engineering 工程
engineering drawing 工程图
engineering manual 工程手册，设计手册
engineering plastics 工程塑料，工程塑胶
engineering strain 工程应变
engineering stress 工程应力
engine-turning lathe 机动车床
engobe 釉底料

engrain 使根深蒂固，把……染成木纹色
engraving 釉底料刻花，刻印
engraving machines 雕刻机
ennea- 【构词成分】九
enneagon 九角形，九边形
enol 烯醇
enol ester 烯醇酯
enol ether 烯醇醚
enolization 烯醇化（作用）
enolphosphopyruvate 磷酸烯醇式丙酮酸
enoyl- 【构词成分】烯酰（基）
enrich 使浓缩［集］
enrichment 浓缩［集，化］（作用），加浓，富集
entire 全部的，整个的，边缘光滑的
entrainer 夹带剂（形成共沸混合物的溶剂）
entrainment 带去，输送，传输，雾沫，夹杂，（气流）带走物
entrainment trap 雾沫分离器，捕沫器
entrap 诱捕［陷］，俘获，捕捉［获］，收集，阻挡，夹［裹］住
entrapped slag 夹渣
entrapment 诱捕，圈套，截留
envelope 封皮，包膜
envelope separator 包状隔板
environmental 环境的，周围的
environmental protection batteries 环保电池
environmental protection 环境保护
environmental stress cracking test 环境应力龟裂试验
environmentally friendly 对环境友好的，保护生态环境的，对生态环境无害的
enyne 烯炔
enzyme electrode 酶电极
epitaxial 外延的，取向附生的
epitaxial growth 叠晶生长，同轴生长（晶体），外延生长
epibond 环氧树脂类黏合剂
epikote 环氧类树脂

epilamens 油膜的表面活性
epi-layer 外延层
epinephelos 混浊的
epiplasma 超等离子体
epipolic 荧光性的
episome 附加体，游离基因，游离体
episulfide 环硫化物
epitaxial 外延的，取向附生的
epitaxis 外延生长，晶体定向生长
epitaxy 外延，取向附生，外延附生
epithermal 超热［能］的，浅层［低温］热液的
epoxidation 环氧化（作用）
epoxide 环氧化物
epoxy 环氧的；环氧树脂
epoxy coating 环氧树脂喷漆；环氧涂层
epoxy plastics 环氧塑料
epoxy resin paint 环氧树脂漆
epoxy resin 环氧树脂
epoxyethane 环氧乙烷
equalizing 平衡，补偿；均衡的
equalizing charge 均衡充电
equalizing filter 均衡滤波器，平衡滤波器
equation 等式，方程
equatorial bond 平（伏）键
equiaxed 由等轴晶粒组成的，各方等大的
equiaxial 等轴的
equiaxial grain 等轴晶粒
equicohesive 等强度的，等内聚的
equidimension 等尺寸，同大小
equidirectional 同向的
equidistant 等距的，距离相等的
equidistant surface 等距曲面
equidistributed 等分布的
equidistribution 均匀分布，等分布
equifrequency 等频（率）
equigranular 等粒度的，均匀颗粒的，同样大小（颗粒）的
equilibrant 平衡，平衡力，平衡状态
equilibrium 均衡，平静，平衡（状态，性，图，曲线）

equilibrium condition 平衡条件
equilibrium constant 平衡常数
equilibrium diagram 平衡图，合金相图
equilibrium electrode potential 平衡电极电位
equilibrium potential 平衡电位
equilibrium reversible potential 平衡可逆电位
equilibrium segregation 平衡偏析
equilibrium segregation coefficient 平衡偏析系数
equilibrium state 平衡状态
equipartition 均分，矩形分布，平均分配
equipment 器材，设备
equipment first inspection 设备首检
equipment for surface treatment 表面处理工具
equipotential volume 等电势（位）体
equivalent 等价物，相等物；等价的，相等的，同意义的，当量的
equivalent circuit 等效电路
equivalent conductance 当量电导
equivalent conductivity 当量电导率
equivalent point 等当点
eradiate 辐射，发射
eradicate 根除，根绝，消灭
erasability 可擦度，可擦性
erasable 可消除的，可抹去的，可删除的
erase 抹去，擦除
erasibility 耐擦性（能）
erection 安装，建立，竖［直］立
eremacausis 慢性氧化，腐败，缓慢氧化
Erichsen test 拉伸性能试验，杯突试验，埃里克森试验
eridite 电镀中间抛光液
eriometer 衍射测微器，微粒直径测定器
erlenmeyer flask 锥形瓶
erode 腐蚀，侵蚀，受腐蚀
erode grain 蚀纹
eroded 被侵蚀的，损坏了
erodent 腐蚀剂；侵蚀的

erodible 受过腐蚀的，易受侵蚀的
erose 啮蚀状的，凹凸不平的，不规则形状的
erosion 侵蚀，腐蚀［浸］，剥蚀，磨蚀，冲蚀，风化
erosion-corrosion 磨耗腐蚀，冲击腐蚀
erosion-resistance 耐冲蚀性
erosive 腐蚀的，冲蚀的，侵蚀性的
erratic 不稳定的，非定期的，不规律的，漂游的，漂移的，杂散的
erratic flow 涡流，扰动流
escalate 逐步增强，逐步升高，使逐步上升
essential 本质，要素，要点；基本的，必要的，本质的，精华的
essential value 酯化值
essential parameter 基本参数
ester 酯
ester value 酯化值
esterification 酯化（作用）
esterify （使）酯化
esterlysis 酯解作用
esters solvent 酯系溶剂，酯类溶剂
estimate 估量
estimated 估计的
estimating 估计，估量，预算
estimation 估计
etch 刻蚀，腐蚀剂，蚀刻
etch back 回蚀．深蚀刻
etch bands 侵蚀带
etch fiqure 蚀迹
etch pit density 腐蚀坑密度
etch rate 蚀刻速率
etch residue 蚀刻残余物
etch selectivity raito 蚀刻选择比，蚀刻选择性
etch uniformity 蚀刻均质性
etchant 侵蚀液，蚀刻液
etching 蚀刻，侵蚀
etching machine 蚀刻机
etching primer 腐蚀底漆，磷化底漆，活性底漆
eteline 四氯乙烯
ethamine 乙胺
ethanal 乙醛
ethanamide 乙酰胺
ethane 乙烷
ethanol 乙醇，酒精
ethanolysis 乙醇解（乙醇分解）
ethene 乙烯
ether 醚，醚基
ethers solvent 醚系熔剂
ethide 乙基
ethidene 亚乙基
ethoxy 乙氧基的
ethoxyl 羟乙基，乙氧基
ethoxylation 乙氧基化
ethyl 乙基
ethyl acetate 乙酸乙酯
ethyl alcohol 乙醇
ethyl ether 乙醚
ethylamine 乙胺，氨基乙烷
ethylate 乙醇化物，乙醇盐，使引入乙基类
ethylation 乙基化
ethylene 乙烯
ethylene glycol 乙二醇
ethylene perchloride paint 过氯乙烯漆
ethylenediamine tetraacetic acid（EDTA） 乙二胺四乙酸
ethylenediamine 乙二胺
europia 氧化铕，三氧化二铕
europium 铕
eutectic 共晶，共熔的，容易溶解的
eutectic alloy 共晶合金，低共熔合金
eutectic bonding 共晶接合
eutectic carbide 共晶碳化物
eutectic cast iron 共晶铸铁
eutectic mixture 共晶混合物，低熔混合物
eutectic phase 共晶相
eutectic reaction 共晶反应
eutectic structure 共晶组织

eutecticum 共晶
eutectiferous 低共熔混合物类型的，（亚）共晶（体）的
eutectiform 共晶状
eutectoid 类低共熔体；共析的
eutectoid alloy 共析合金
eutectoid composition 共析成分
eutectoid hardening 共析硬化
eutectoid line 共析线
eutectoid point 共析点
eutectoid reaction 共析反应
eutectoid structure 共析组织
eutectoid temperature 共析温度
eutectoid transformation 共析变态
eutectometer 快速相变测定仪，凝点记录仪
eutexia 易熔性，稳定状态，稳定结核性，低共熔性
eutropic 异序同晶的
eutropy 异序同晶现象
evacuate 抽空，撤退，排泄［空］
evacuated 抽空的
evacuated chamber 真空容器，真空室
evacuation 抽空，排空
Evans diagram 腐蚀极化图，埃文斯图
evanescent 易消散的，逐渐消失的，无限小的，短暂的，不稳定的，易挥发的
evanescent wave 损耗波
evaporability 挥发性，可蒸发性，汽化性
evaporable 易蒸发性的，易挥发的
evaporant 蒸发剂，蒸发物
evaporate 蒸发
evaporated 蒸发的
evaporating dish small 蒸发皿
evaporation 蒸发
evaporation coefficient 蒸发系数
evaporation loss 蒸发损失
evaporation material 蒸发材料
evaporation ratio 蒸发比
evaporation source 蒸发源
evaporativity 蒸发率，蒸发度
even 平坦［滑］的，有规律的，不变的；使均匀（平衡，平均）
even cooling 均匀冷却
evermoist 常湿［潮］的
evolute 渐屈线；渐屈的
evorsion 涡流侵蚀
ex situ 非现场，非原位，天然状态外
exact 准确的，精密的，精确的
exaggerate 使扩大，使增大，夸大
exaltation 使提升，升高，纯化，精炼
examination 试验，检验
examination area 检验范围，检测范围，监测区域，测试区域
examination region 检验区域，检测范围
examined 检验过的，试验过的
excentric 不同圆心的，离心的，偏心的
excess 过量；过量的，额外的
excessive defects 过多的缺陷
excessive gap 间隙过大
exchange 交换，替换
exchange capacity 交换容量
exchange current 交换电流
exchange current density 交换电流密度
exchange effect 交换作用
exchange resin 离子交换树脂，交换树脂
excitation 激发，激励
excitation spectrum 激发光谱
excited state 激发态
exergic 放能的，放热的
exergonic 释出能的，放出能的
exesion 腐蚀
exfoliate 使片状脱落，使呈鳞片状脱落，片状剥落，鳞片样脱皮
exfoliation 脱落，脱落物，表皮剥脱，页状剥落
exfoliation corrosion 剥蚀
exfoliation-adsorption 剥离吸附法
exhalant 发散管，蒸发管；呼出的，蒸发的，发散的
exhale 呼气，发出，发散，使蒸发
exhaust 排出口，用［耗］尽，排出［气］，排气装置；排出，耗尽

exhaust air washing system　排气洗净装置
exhaust pipe　排气管
exhaust pressure　排气压力
exhaust tube　排气管
exhausted resin　失效树脂
exhaustion　失效，耗尽
exhaustion creep　耗竭潜变
exhaustive　彻底的，消耗的，抽〔排〕气的
exhibit　表现，陈列，呈现
exigible　可要求的
exiguity　微小，些许，稀少
exiguous　稀少的，细小的
exility　稀少，纤细，稀薄
exo-　【构词成分】外，在外，外面的，支，放
exodic　离心的，传出的，输出的
exoergic　放能的，放热的
exogenous　外生的，外成的，外因的
exogenous metallic inclusion　外来金属夹杂物
exogenetic　外生的，外因的，外源性的
exophytic　外部生长的，外生型的，向外生长的
exorbitance　过度，不当
exosmic　外渗的，渗入的
exosmose　外渗现象，外渗
exotherm　放热曲线，（因释放化学能而引起的）温升
exothermic　发热的，放出热量的
exothermic reaction　放热反应
exothermicity　放热性
expansion　膨胀，阐述，扩张物
expanded　膨胀的，扩大的，延伸的；使……变大，扩大，伸展
expansion coefficient　膨胀系数
expansion crack　膨胀裂缝，膨胀裂纹〔裂痕〕
expansion dwg　展开图
expansion joint　伸缩接头，补偿器
expedance　负阻抗

experimental　实验的
explosive　爆炸的，爆炸性的，爆发性的
explosive evaporation　沸腾蒸发
exponential　指数，例子；指数的，幂数的
expose　揭露，揭发，使曝光，显示
exposed　裸露的，暴露的；揭露，暴露
exposure　暴露，揭发，公开
exposure chart　曝光曲线
exposure test　暴露实验
exsiccant　干燥剂，脱水剂；能除湿的
exsiccate　使干燥，变干
exsiccation　脱水，干燥，干燥作用，干燥法
exsiccative　除湿的，使干燥的
exsiccator　除湿剂，除湿器
exsolution　出溶作用，脱溶
exsudation　渗出，盐剥作用
extend　延伸，扩大，推广，伸出，给予，使竭尽全力，对……估价
extended dislocation　扩展位错，延伸位错
extended X-ray absorption fine structure（EXAFS）　X射线精细结构
extender filler　填料
extender pigment　体质颜料
extension　延长，扩大，伸展
extensional　外延的，具体的，事实的
extensional vibration　纵向振荡，拉伸振动，扩张振动
extensionality　外延性
extensive　广泛的，大量的，广阔的
extensive repair　大修，大修理
extensometry　应变测定，伸长测定
extent　扩大，延伸
extenuate　减轻，低估，降低，衰减
exteriorly　从外表上看，从外部
external　外部，外观，外面；外部的，表面的
external cause　外因
external conversion　外转换
external indicator　外指示剂
external plasticization　外部增塑化

external reflection mode 外反射模式
external standardization 外标法
extinction 消失，消灭，废止
extinction coefficient 消光［声］系数
extract 提取物，萃取，提取
extractability 可萃取性，提取性
extractable 可抽出的，可榨出的，可推断出的
extractant 提取剂，提炼物，分馏物，萃取剂
extracted 提取的
extractibility 可萃取性
extracting 提取
extraction 提取法，萃取法
extraction voltage 引出电压
extractive 提取物，抽出物；提取的，可萃取的
extractive metallurgy 提取冶金
extractor 提取器，萃取器
extraneous 外来的，没有关联的，局部的，附加的，无的，不重要的
extraordinarily 非常，格外地，非凡地
extraordinarily high temperature 超高温
extrapolation 外推法，推断

extra-pure 极纯的，特纯的，高纯度的
extreme 极端，末端，最大程度；极端的，极度的
extrinsic 非本质的，外在的，外部的，外来的，非固有的
extrinsic semiconductor 含杂质半导体
extrudability 可挤压性
extrude 挤出，压出，使突出，逐出，突出，喷出
extruded 压出的，受挤压的
extruding 挤压
extrusion molding 挤出成形
extrusion 挤［推，喷，赶］出
exudate 分泌液，流［渗］出物
exudation 渗出，渗［流］出物，热［熔，烧］析
exuding 缓慢流［渗，分泌］出
exutory 取出，流出，脱除剂，诱导剂；诱导的
eye survey 目测
eyelet 小孔，针眼，在……上打小孔，给……嵌入镶孔金属环
eyepies 目镜

F

fabric 织物，布，组织，构造，建筑物
fabrication 加工，制造
fabrication drawing 制造图纸，制作图
fabrication holes （印刷电路板上的）工艺孔
face guard 护面具，面罩
face perpendicular cut 垂直面切割
face-centered 面心的
face-centered cubic (FCC) 面心立方
face-centered orthorhombic 面心斜方
facing 饰面，刮面，表面加工；面对的，盖面的，外部的
factor 系数
fading 褪色，枯萎，衰退；使褪去，使消失
failure 故障，失效
failure analysis 失效分析
failure mechanism 失效机理
failure mode 失效模式
failure mode effectiveness 失效模式分析
failure wear 故障磨损
faint 微弱的，轻微的，暗［淡］的；消失，变弱［淡］
faint red 淡红色
faints 劣质酒精
fair 平直的，（外形）平顺的；把……做成流线型，使流线型化，整流，修整，整形
faired （做成）流线型的，整流的，流线型的，减阻的
fairing 整流罩，流线型；光顺的，减阻的
fall off 减少，跌落，变坏，脱落
fallback 低效率运行
falling ball impact test 落球冲击试验
falling weight test 落重试验
false bedding 假［不规则］层理，交错层
false indication 假指示，虚假指示

family 族，系列
fan 风机
fan conveyer 旋转式运送机
fan drift 通风道
fanned 扇形的，带翼的
fanning 铺开，展开，形成气流；吹风
fanning strip 扇形板［片］
far field 远场
far-infrared 远红外（线）的
faradaic 感应电流的，法拉第的
faradaic process 法拉第过程
faradaic rectification 法拉第整流法
faraday 法拉第
Faraday's law 法拉第定律
faradic 感应电的，法拉第的
faradic current 法拉第电流，感应电流
faradic impedance 法拉第阻抗
fast atom bombardment mass spectrometry (FABMS) 快原子轰击质谱
fast fourier transformation 快速傅里叶变换
fast neutron detectors 快中子探测
fast plating 快速电镀
fast screen 短余辉荧光屏
fast yellow 耐晒黄
fastening 扣紧，紧固
fastening motion 夹紧
fat and oil 油脂
fat hardening 油脂硬化
fat hydrolyzing process 油脂水解法
fat solvent 油脂溶剂
fatigue 疲劳
fatigue corrosion 疲劳腐蚀
fatigue crack 疲劳裂痕
fatigue fracture 疲劳断裂
fatigue life 疲劳寿命
fatigue limit 疲劳限度，疲劳极限

fatigue machine 疲劳试验机
fatigue strength 疲劳强度
fatigue test 疲劳试验
fatigue wear 疲劳磨损
fatty 脂肪的，肥胖的，多脂肪的，脂肪过多的
fatty acid 脂肪酸
fatty acid anhydride 脂肪酸酐
fatty acid ester 脂肪酸酯
fatty alcohol 脂肪醇
fatty alcohd-polyoxyethlene ether 脂肪醇聚氧乙烯醚
fault 缺陷
fault analysis 缺陷分析
fault indicator 故障指示器，探伤机
faulty 有错误的，有缺点的，出故障的，不合格的，报废的
favorable 有利的
Feal 费尔合金（25％Al，其余为Fe）
feasibility 可行性
feather marking 羽状痕迹
feathery 柔软如羽毛的，生有羽毛的
feathery needles 羽毛针状体
feed 供料
feedback 反馈，回流
feeder 送料机
feeder apparatus 送料［加料，馈电］装置
feeding 供料的
feeding current 馈电电流
feeler 触角［点］，探针，测头，探测器
Fehling's reagent 斐林试剂
fermentation 发酵
Fermi energy 费米能
Fermi level 费米能级
Fermi surface 费米面
ferment 酶，酵素，使发酵
ferrate 铁酸盐，高铁酸盐
ferrated 含铁的，加铁的
ferreous 铁的，含铁的，硬如铁的
ferri- 【构词成分】含（正）铁的

ferric 铁的，三价铁的，含铁的
ferric acid 高铁酸
ferric ammonium sulfate 硫酸铁铵
ferric chloride 氯化铁
ferric compound 正铁化合物
ferric nitrate 硝酸铁
ferric oxide 氧化铁，三氧化二铁
ferricyanate 铁氰酸盐
ferricyanide 铁氰化物
ferriferous 含铁的，含有三价铁的
ferrigluconate 葡糖酸高铁盐
ferrimagnetism 亚铁磁性
ferrous material 铁系金属
ferrite （正）铁酸盐，铁素体，铁氧体
ferritic 铁素体的，铁氧体的
ferritic malleable 铁素体可锻铸铁
ferritic stainless steel 铁素体不锈钢
ferritization 褐铁矿化，铁氧体化
ferritize 铁氧体化
ferro- 【构词成分】表示"含［亚，二价］铁的，铁合金的"
ferroalloy 铁合金
ferrochrome 铬铁合金
ferrochromium 铬铁
ferrocyanide 亚铁氰化物
ferroelectric 铁电性；铁电的
ferroelectrics 铁电体；铁电材料
ferroferric compound 亚铁正铁化合物
ferroferric oxide 四氧化三铁
ferromagnet 铁磁物质，铁磁体
ferromagnetic 铁磁体；铁磁的，强磁的
ferromagnetic substance 铁磁物质
ferromagnetism 铁磁性
ferromanganese 锰铁
ferromolybdenum 钼铁
ferronickel 镍铁
ferrosilicon 硅铁
ferrous 铁的，含铁的，亚铁的
ferrous alloy 铁合金
ferrous chloride 氯化亚铁

ferrous iodide 碘化亚铁
ferrous lactate 乳酸亚铁
ferrous metal 黑色金属
ferrous oxalate 草酸亚铁
ferrous salt 亚铁盐
ferrovanadium 钒铁
ferroxcube 立方结构的铁氧体，铁氧体软磁性材料
ferroxyl 铁锈
ferroxyl indicator 铁锈指示剂
ferroxyl test 孔隙度试剂试验，滤纸斑点试验
ferruginosity 含铁性
ferruginous 铁的，铁锈的，含铁的
ferruginous dross 含铁浮渣
ferrum 铁
ferrum reductum 还原铁
ferry 渡船，摆渡，渡口；（乘渡船）渡过，用渡船运送，空运
fervent 强烈的，炽热的
fetch 取来，接来，拿，取物
fettle 修补［炉］，用炉渣等涂（炉床），涂衬炉床
fettle material 补炉材料
fettling 涂炉床材料，铸件清理，炉底修补
fiber 纤维
fiber stress 纤维强度，纤维应力
fiber structure 纤维结构
fiber texture 纤维组织
fiber toughening 纤维增韧
fiberglass reinforced plastics (FRP) 玻璃纤维增强树脂
fiber-reinforced composite 纤维增强复合材料
fiber-reinforcement 纤维增强
fiberized 纤维化，成纤维，使……成纤维，使……纤维化
fibestos 醋酸纤维，一种乙酸纤维素，塑胶
fibration 纤维化，纤维形成

fibre 纤维，纤维制品
fibreless 无纤维的
fibre-optic 光导纤维的
fibriform 纤维状的，根毛状的
fibrillar 纤维状的，纤丝状的，根毛的
fibrinous 含纤维的，由纤维素形成的，有纤维素的
fibro- 【构词成分】纤维
fibroid 纤维性的，纤维状的，由纤维组成的
fibroillary 微丝的
fibrous 纤维的，纤维性的，纤维状的
fibrous insulation 纤维绝缘，纤维隔热［离］层
fibrous material 纤维材料
fibrous region 纤维区
fickle 易［多］变的，不专的，变幻无常的
fickle colour pattern 变幻无常的色彩模式，不规则彩色图形
Fick's first law 菲克第一定律
Fick's second law 菲克第二定律
fictile 塑造的，陶土制的，可塑造性的
fictitious 虚构的，假想的，编造的，假装的
fictive 想象的，虚构的
fictive temperature 假想温度
fiddling 无用的，无足轻重的，微不足道的
fidelity 保真度，精确
fiducial 基准的
fiducial temperature 基准温度
field 领域，区域，范围
field density 场密度，场强度
field desorption 场解吸法
field electron microscope 场电子显微镜
field emission 场致发射，电场发射，冷发射
field fabricated 工地制造的，现场装配的
field induced migration 场诱导迁移
field installation 现场安装

field intensity 场强度
field ionization 场电离
field of force 力场
figurable 能成形的，可以定形的，可具有一定形状的
figural 用形状表示的，有［成，造］形的
figuration 成［外，定］形，轮廓，图案装饰法
figurative 造型的，赋形的，用图形表现的
figuratrix 特征表面
figure 图，外形，轮廓
figured 有形状的，用图画表现的，有图案的
filaceous 丝状的，含丝的
filament 灯丝，细丝，细线
filamentary 纤维的，似丝的，细丝的，纤细的
filamentary cathode 线状阴极，直热式阴极
filamentation 丝状形成，光丝的形成，（束流的）丝化现象
filamented 有细丝的
filar 丝的，丝状的，线的
file 锉刀
filiform 丝状的，纤维状的
filiform corrosion 丝状腐蚀
filiform corrosion test 丝状腐蚀性试验
filing 锉削加工
fill factor 填充系数
fill in 填写
filler 充填剂，填充物，填充料
filler metal 熔填金属
filler rod 焊条
filler speak 填充料斑
fillet 镶，嵌，倒角，凸起；用带缚，加边线，修圆，把……切成片
filling 填
filling water test 充水试验
film applicator 涂膜涂布器，刮膜机

film base 片基，胶片基
film coefficient 膜系数
film condensation 薄膜冷凝
film former 成膜物，成膜剂
film forming ability 成膜能力
film forming ingredient 涂膜主要成分
film forming matter 成膜物，成膜物质
film thickness gauge 漆膜厚度计，膜厚计
filter 滤波器，过滤器，筛选，滤光器；过滤，渗透，用过滤法除去，滤过，渗入
filter aid 助滤剂，助滤器
filter bed 过滤层，滤床，滤水池
filter cake 滤饼
filter cake washing 滤饼洗涤
filter cloth 滤布
filter cone 滤斗，锥形滤器
filter element 过滤芯，滤波元件
filter gauze 滤布，滤网
filter layer 过滤［渗透，透水］层
filter liquor 滤液
filter medium 过滤介质，滤材，滤器填料
filter paper 滤纸
filter press 压力过滤器，压滤机
filter pump 滤泵
filter stand 漏斗架
filter stick 滤棒
filter support 漏斗架
filter tube 滤管
filtering 过滤
filtering basin 过滤池
filtering media 滤料
filtering surface 过滤面，滤面
filtrate 滤液；过滤，筛选
filtration 过滤，筛选
filtration equipment with coagulation 凝聚过滤设备
filtration screen 滤网
final 最终的
final product 最终产品，最后产物
final voltage 终止电压，截止电压
fine alignment 精细对准

fine chemistry　精细化学
fine blanking　精密下料加工
fine fissure　细裂纹
fine grading　细级配
fine grain deposits　微晶沉积
fine grained　细纹理，细粒性分布；细粒（度）的，小碎块的
fine particle　细颗粒
fine pearlite　细片状珠光体
fine powder　细粉，超微粒
fine structure　精细结构
fineness　细度
fineness gauge　细度计，细度测定计
fines　细料，粉末，细骨料；使精细
fining　澄清，澄清剂
finish　修正，精修，研磨，抛光，涂层，涂料，漆面，抛光剂
finish cut　精加工
finish sanding　修饰性打磨
finishability　饰面难易度，易修整性，精加工性
finished product　成品，最终产品
finishing　精整加工
finishing allowance　加工余量
finishing coat　面漆，最后一道涂工
finishing compound　精饰化合物
finishing machines　修整机
finite diffusion impedance　有限扩散阻抗
finned　有翼的，有散热片的，有加强筋的
finning　加肋，用肋加固
finsen lamp　紫外线灯，水银灯，汞灯
fire extinguisher　灭火器
fire foam　泡沫灭火剂
fire metallurgy　火法冶金
fire point　燃点，着火点
fire proof material　耐火材料
fire resistance　耐火性，抗燃烧性
first coat　底涂，第一涂层
first dislocation generation　初次发生位错
first stage annealing　第一段退火
first inspection　首检

first stage tempering　低温回火
firth hardometer　费氏硬度计
firty　耐火性
fish eye　鱼眼，鱼眼状裂纹，白点焊接
fish scale　鳞状破面
fisheye preventer　防鱼眼剂
fiss　分裂，裂变
fiss-　【构词成分】分，裂
fissible　可裂变的，可分裂的
fissile　分裂性的，易分裂的
fissility　易裂性，可裂变性
fission　分裂，裂变
fissionability　可裂变性，裂变度，裂变能力
fissionable　可裂变物质；可分裂的，可作核裂变的
fissura　裂隙，裂纹
fissuration　形成裂隙，龟裂，裂开
fissure　裂纹，裂隙；裂开
fissuring　裂隙，节理
fit tolerance　配合公差
fitful　不规则的，间歇的
fitment　设备，配件，附件
fitting　配件
fix oil　硬化［非挥发性，脂肪］油
fixation　固定，定位，定影，不挥发
fixative　固定剂，固着物，定色剂；固定的，防挥发的，定色的，防褪色的
fixative salt　定影剂
fixed　固［确，稳，恒］定的，不变［动］的，定位的，凝固的，不易挥发的
fixed acid　不挥发酸
fixed air　不流动空气
fixed bed　固定床
fixed bed ion exchange　固定床离子交换
fixed resistance discharge　定阻抗放电
fixedness　固定，稳固，不变，硬度，刚性，凝固［稳定，耐挥发］性
fixing　固定，夹紧，嵌固，安装，修理，整顿
fixture　固定装置，夹具

flake 小［薄，鳞，絮］片，片状粉末，絮状体，絮团，裂纹；使（成片，分层）剥落，去氧化皮
flake graphite 片状石墨
flakiness 片状，成薄片，片层分裂
flaking 表面（片状，分层）剥落，去氧化皮；易剥落的
flaky 薄片的，薄而易剥落的
flaky crystal 片状结晶
flaky grain 片状颗粒
flaky material 片状材料
flaky resin 片状树脂
flame 火焰；燃烧，发火焰，加热于，闪耀，发光，照亮
flame ablation 溶化烧蚀，熔化烧蚀
flame analysis 火焰分析
flame annealing 火焰退火
flame arc 弧焰
flame arc lamp 弧光灯
flame arrester 灭火器
flame chipping 烧剥
flame couple 热电偶
flame current 电弧电流
flame cutting 火焰切，气割
flame damper 灭火器，防火器
flame descaling 火焰除锈，火焰除锈皮
flame hardening 火焰硬化，火焰淬火
flame retardant 耐燃剂，阻燃剂
flame scaling 锌镀层火焰加固处理，热浸镀锌
flame seal galvanizing 火封软熔热镀锌法
flame spectrophotometer 火焰分光光度计
flame strengthening 火焰强化
flame surface quenching 火焰表面淬火
flame treatment 火焰处理
flame-spraying 焰喷
flammability 可燃性
flammable 可燃的，易燃的
flammable liquid 可燃液体，易燃液体
flange 边缘，轮缘，凸缘，法兰
flange connection 凸缘连接
flange gasket 法兰垫片法兰衬垫
flange joint 凸缘接头，凸缘接合
flange wrinkle 凸缘起皱
flanging 凸缘加工，折边，翻口，凸缘
flap 片状物，阻力板，簧片，闸门
flare 闪光，闪耀，耀斑，爆发；闪耀，闪光，燃烧；
flareback 回火，逆火
flared 裂缝呈张开，喇叭口，爆发的，加宽的
flaring 燃烧的，发光的，喇叭形的，张开的，扩口的，漏斗状的
flaser 压扁状的
flaser texture 鳞状组织
flash 闪镀，飞边，溢边
flash arc 闪光电弧
flash baking 快速烘干
flash distillation 快速蒸馏，闪蒸
flash melting 热熔
flash off 晾干
flash plate 闪镀
flash point 闪点
flash rust inhibitors 闪锈抑制剂
flashing 闪烁的，闪光的
flashover 飞弧，击穿，闪络，跳火；产生飞弧
flashover characteristic 闪络特性，放电特征
flashover strength 轰燃强度，火花击穿强度
flashover welding 闪光焊，弧焊
flashy 闪光的，瞬间的
flashy load 瞬间载荷
flat 平整；平坦的，扁平的，浅的；使变平，变平
flat crystal 片状晶体
flat grain 平纹，弦切纹理
flat paint 无光涂料，消光涂料
flatten 变［弄，修，拉，展，锤，整］平，使……平坦，使无光泽
flattener 压平机，压延机，扁条拉模

flattening　整［修，校］平，扁率，压扁作用，补偿
flattening agent　整平剂，（涂料）平光剂
flatting　流平性；使物体平滑的方法，使油漆无光泽的方法
flatting agent　消光剂，平光剂
flavescent　逐渐变黄的，浅黄色的，带黄色的，变黄色的
flaw　刮伤，缺陷，瑕疵，裂纹
flaw characterization　伤特性，缺陷特征
flaw detection　探伤检验，缺陷检验
flaw detector　探伤仪，裂痕探测仪
flaw echo　缺陷回波，缺陷的回波信号，缺陷回声
flawless　完美的，无瑕疵的，无裂缝的
flawmeter　探伤仪；探伤
flaxen　淡黄色的，亚麻的，亚麻色的
flaxseed oil　亚麻油，麻子油
fleck　斑点，微粒，小片；使起斑点，使有斑驳
fleckless　无斑点的，洁白的
fleur　粉状填料，粉状填充物
flex　屈曲；折曲，使收缩
flex crack　屈挠龟裂
flexibility　弹性，可弯曲性，柔顺性
flexibility characterist　柔度特性
flexibilizer　增韧剂，增塑剂
flexible　灵活的，柔韧的，易弯曲的
flexible rigidity　弯曲刚性
flexion　弯曲，弯曲状态，弯曲部分
flexional　可弯曲的
flexivity　挠度，曲率
flexuose　弯曲的，曲折的，波状的，锯齿状的
flexuosity　弯曲性，弯曲状态，波状
flexural　弯曲的，曲折的，挠性的
flexural center　弯曲中心
flexural rigidity　抗弯刚度，抗挠刚度
flexural strain　弯曲应变，挠曲应变
flexural strength　抗弯强度
flexural stress　弯曲应力，挠曲应力

flexure　屈曲，弯曲部分，打褶，挠曲
flick　污点，斑点；轻打，弹动
flight　飞行［驰，逝，航，程］，行程
flighted dryer　淋式干燥器
flimsy　薄纸；易损坏的
flinty　坚硬的，极坚硬的，强硬的
float　（使）浮动，（使）漂浮，自由浮动
float polishing　浮动抛光
float valve　浮球阀
floatability　浮动性，漂浮性，可浮选性
floatable　可漂浮的，可浮选的
floatation　浮选，漂浮性
floatation oil　浮选油
floating　流动的，漂浮的，浮动的
floating agent　浮选剂
floatstone　浮石
floc　絮凝物，絮状物
floc point　絮凝点
flocculability　絮凝性
flocculable　可絮凝的，易絮凝的
flocculant　絮凝剂，凝聚剂
floccular　絮片的，绒球的，絮凝的
flocculate　絮凝，絮结
flocculated structure　絮凝结构
flocculating tank　絮凝池
flocculation　絮凝，凝聚，絮状沉淀法
flocculation agent　絮凝剂
flocculation aid　助凝剂
flocculation value　凝结值
flocculator　絮凝器
floccule　絮状物，絮凝粒
flocculence　絮凝性，絮凝状
flocculent　絮凝剂；絮凝的，绒聚的，绒毛状的，含絮状物的
flocculent deposit　絮凝状沉淀，凝絮状沉积
flocculent structure　絮凝结构
flocculus　絮凝，绒球，絮片，絮状物
flood spray pretreatment system　溢流喷射式前处理
flooding　注水，浸渍，溢流，满溢，（油

漆干燥时或加热时）变色
flooding pipe 溢流管
flooding value 溢流阀
floss 絮状物，（浮于熔化金属表面的）浮渣
flotation 浮选
flotation agent 浮选剂
flour 粉状物质；粉末乳化，研成粉末
flour filler 细粉状填料
flow 流动，流量；淹没，溢过
flow ability 流动性
flow agent 流平剂
flow coating 流型涂膜
flow injection analysis 流动注射分析法
flow mark 流痕
flow meter 流量计
flow off 溢流口
flow sheet 流程图，作业图
flow through 溢流道
flowage 流动，泛滥，流出
flowchart 流程图，作业图
flowing 流动的，平滑的，上涨的
flowing furnace 熔化炉
flowline 流线，气流线，晶粒滑移线
flowmeter 流量计，流速测定仪
flowrate 流量，流速
fluorescent 荧光的
fluorescent lamp 荧光灯
flucticulus 波纹，波痕，微波
fluctuant 变动的，波动的，起伏的
fluctuate 波动，涨落；使波动
fluctuation 起伏，波动
flue 烟道，送气管，通气道
flue gas 烟道气，废气
fluent 流畅的，液态的，畅流的
fluent metal 液态金属
fluffy 蓬松的，松软的，绒毛的，易碎的
fluid 流体；流动的，流畅的，不固定的
fluid characteristics 流体特性
fluidal 流体的
fluidal structure 流状构造，流纹构造

fluidic 流体的，射流的，能流动的
fluidification 液化，流化
fluidify 使成为流动，使流体化，液化，积满液体
fluidity 流体性
fluidity index 流体性指数
fluidization 流体化，流化作用
fluidize 使流体化，使液化
fluidized bed 流体化床
fluo- 【构词成分】氟，荧（光）
fluoborate 氟硼酸盐
fluoration 氟化，氟化作用
fluorescence 荧光
fluorescence analysis 荧光分析
fluorescence efficiency 荧光效率
fluorescence life time 荧光寿命
fluorescence efficiency 荧光效率
fluorescence quantum yield 荧光量子产率
fluorescence quemching method 荧光熄灭法
fluorescence spectrum 荧光光谱
fluorescent 荧光的，发亮的
fluorescent additives 荧光剂
fluorescent examination method 荧光检验法
fluorescent magnetic powder 荧光磁粉
fluorescent material 荧光材料
fluorescent penetrant 荧光渗透剂
fluorescent penetrant inspection 荧光渗透检验
fluorescent substance 荧光物质
fluorescent X rays 荧光 X 射线，特性伦琴射线
fluorhydric acid 氟酸，氢氟酸
fluoric plastic 氟塑料
fluoridate 向……加入氟化物
fluoridation 氟化作用［反应］，加氟作用
fluoride 氟化物
fluoridize 用氟化物处理，涂氟，氟化
fluorimetric 荧光的
fluorimetry 荧光分析，荧光测定法

fluorinate 使与氟化合，用氟处理
fluorinate 氟化的
fluorination 氟化（作用，法）
fluorine 氟
fluorite 萤石，氟石
fluorizate 氟化
fluorization 氟化作用
fluoro- 【构词成分】氟（代），荧光
fluoroalkylpolysiloxane 氟烷基聚硅氧烷
fluorocarbon 碳氟化合物
fluorochemical 含氟化合物，氟化物
fluorochlorohydrocarbon 氟氯烃
fluorochrome 荧色物，荧光染料
fluoroelastomer 氟橡胶，含氟弹性体
fluorography 荧光照相术
fluorohafnate 氟铪酸盐
fluorol 氟化钠
fluorolube （含氧设备用）氟碳润滑剂
fluorolubricant 氟化碳润滑油
fluorometric 荧光的，使用荧光测定法的，使用荧光计的
fluorometry 荧光分析法
fluorophotomeric detector (FD) 荧光检测器
fluoroscopical 荧光镜透视检查的，X射线透视检查的
fluoroscopy 荧光检查法
fluorosurfactant 含氟表面活性剂
fluorous 氟的
fluosilicate 氟硅酸盐
fluosilicic acid 氟硅酸
fluosolids 流化层，沸腾层
fluostannate 氟锡酸盐
fluotantalate 氟钽酸盐
fluotitanate 氟钛酸盐
fluozirconate 氟锆酸盐
flush （强液体流）冲洗，洗涤，冲刷；倾泻，涌；大量的，丰足的，使齐平，用水冲洗
flushing 冲洗，填缝
flute （凹）槽，（刃）沟，波纹；波纹的

flute profile 槽形，沟形
fluted 有凹槽的，带槽（纹）的，有波纹的
fluting 开［切，刻凹］槽，弯折，折断，折纹
flutter 摆动；拍，飘动
flux 流量，通量，熔化，流出，使熔融，用焊剂处理，助熔剂
flux leakage field 磁通泄漏场，漏磁场
fluxibility 助熔性，熔度
fluxible 可熔的，可流动的，可熔解的，易熔的
fluxion 流动，溢出，熔解，流动
fluxion structure 流状构造，流纹构造
fluxmeter 磁通计，磁通表，磁通量计
fly ash 飞灰
flyspeck 污点，小斑，黑斑
foam 泡沫；海绵状的；起泡沫
foam control agent 泡沫控制剂
foam drainage 泡沫排液
foam glue 泡沫胶粘剂
foam persistence 泡沫持久性
foam polystyrene 泡沫聚苯乙烯
foam separation 泡沫分离
foam separator 泡沫分离器
foam stabilizer 泡沫稳定剂
foam breaker 消泡剂
foamability 起泡性，发泡性，发泡能力
foamable 发泡的，会起泡沫的
foamer 发泡剂，泡沫发生器
foaming 起泡，发泡
foaming test 起泡试验
foamite 泡沫灭火剂
foamless 无泡沫的
foamy 泡沫的，起泡沫的，全是泡沫的
focal 焦点的，在焦点上的
focal distance 焦距
focalization 集中焦点，焦距调整
focalize 使集中在一点上，调节……的焦距，聚焦，限制于小区域
focus 焦点，中心，清晰，焦距；使集中，

使聚焦
focusing 聚焦，调焦
focusing probe 聚焦探头
fog lubrication 油雾润滑，喷雾润滑
fog quenching 喷雾淬火
fogging 成雾，蒙上水汽
foggy 有雾的，模糊的，朦胧的
fold 褶痕
foil 箔，薄金属片；以箔为……衬底
foil insulation 箔绝缘
folding 可折叠的
foliation 剥层腐蚀
foliose 多叶的，叶子覆盖的
foraminate 有小孔的，多孔的
foraminiferous 带孔的，有孔的
foraminulate 有小孔的
forbidden band 禁带
forbidden transition 禁阻跃迁
forced air cooling 强制空气冷却
forced convection 强制对流
forced draft 强制通风
forced drying 强制干燥
forced gas convection 强制气对流
forced vibration 强制振动
fore- 【构词成分】前，先，预
forecooling 预冷
foreign body 夹杂物，外来物，杂质
forepump 预抽泵，前级泵
forepump system 前级泵系统，前级抽空系统
forestage 前级的
forevacuum 前级真空
forewarmer 预热器
forge 锻造
forge iron 锻铁
forge hot 热锻
forge press 锻压机
forge welding 锻造熔接，锻接
forgeability 锻造性，可锻性
forgeable 可锻造的，延性的
forged 锻的，锻造的

forged steel 锻钢
forging 锻（件），锻造
forging die 锻模
forging roll 轧锻
fork 叉子，分叉；做成叉形
forked 有叉的
forklike 像叉一样的
form 形式，形状［态］，外形，方式，表格；构［组］成，排列，组织，产生
formability 成形性，可模锻性
formable 可成形的，适于模锻的
formaldehyde 甲醛
formaldehyde solution 甲醛溶液，福尔马林
formale 聚乙烯
formale copper wire 聚乙烯（绝缘）铜线
formate 甲酸盐，蚁酸盐
formation 形成，构造，编队
formic 甲酸的，蚁的，蚁酸的
formic acid 甲酸
formica 胶木，配制绝缘材料，热塑性塑料
forming 形［冶，组，构，编］成，仿［变］形，成型，模锻，冲压，模压，翻砂
forming die 成型模，精整模
formimino- 【构词成分】亚胺甲基
formiminoether 亚胺甲基醚
formless 没有形状的，无定形的
formoxy- 【构词成分】醛氧基
formoxyl- 【构词成分】甲酰基
formpiston 模塞，阳模
formvar 聚醋酸甲基乙烯酯
formyl 甲酰，甲酸基
formylate 甲酰化
formylic acid 甲酸，蚁酸
fornicate 弓形的，拱状的，弯曲的，穹隆状的
fortifier 增强剂，加固物，强化物
fortify 加强，增强
forward bias 正向偏压，前向偏移

forward extrusion 顺向挤制，正挤压
fossa 凹，小窝
foul 污物；难闻的，恶臭的，污浊的；弄脏，使……阻塞
fouling control agent 污垢控制剂
fouling factor 污垢系数
fouling index 污染指数
fouling resistance 污垢热阻
founding 铸造，铸体，熔制
founding furnace 熔炉
founding property 铸造性能
foundry 铸造，铸造类
foundry equipment 铸造设备
Fourier transform infrared spectrometer (FTIR) 傅里叶变换红外光谱仪
Fourier transform infrared spectroscopy 傅里叶转换红外光谱学
Fourier transform raman spectrometer 傅里叶变换拉曼光谱仪
four (point) probe method 四（点）探针法
fraction 馏分，碎片，小部分，成［级］分，粒级
fraction void 疏松度
fractional 部分的，分馏的，分级的
fractional crystallization 分段结晶
fractional distillation 分馏
fractional extraction 分馏萃取
fractional neutralization 分级中和
fractional precipitation 分级沉淀
fractionate 分离，使分馏，把……分成几个部分
fractionating column 分馏柱
fractograph 断口组织
fractography 断口形貌学，金属断面的显微镜观察
fracture 断口［面］，裂隙［痕，面］
fracture mechanics 断裂力学
fracture stress 破断应力
fracture surface 破断面，断口
fracture toughness 断裂韧性

fracture transition 脆断转移
fragile 脆的，易碎的
fragment 碎片，片断或不完整部分；使成碎片
fragment ion 碎片离子
fragmental 破片的，碎片的，不完全的，零碎的
fragmentation 破碎，分裂
fragmentize 分裂，裂成碎片
frame filter press 框式压滤机
frame front 边框
frame work structure 构架结构
framed 有构架的，构架的
framing 框架，结构
fray 磨损处；使磨损
fraying 磨损，摩擦后落下的东西，织物磨损后落下的碎片
frazzle 磨破；使磨损，使磨破
freak 奇异的，反常的
freakish 畸形的，奇特的
freckle 斑点，（镀锡薄钢板的缺陷）孔隙；使产生斑点
free acid 游离酸
free alkali 游离碱
free available chlorine 游离有效氯
free carbon 单体碳
free chlorine 游离氯
free convection 自然对流
free electron 自由电子
free energy 自由能
free radical 自由基
free radical reaction 游离基反应
freeze drying 冷冻干燥
freeze drying equipment 冻干机
freeze-thaw stabilizers 冻融稳定剂
freezing 冷凝，冻结凝固；冰冻的
freezing and thawing test 冻结融解试验
freezing cuve 冷凝曲线
freezing method 冷凝法
freezing point 冰点，凝固点
freezing range 凝固范围

freezing resistance 抗冻性，耐冻性
freight 货运，运费，船货；运送，装货，使充满
freon 氟利昂，氟氯烷，二氯二氟甲烷
frequency 频率
frequency constant 频率常数
frequency dispersion 频散
frequency factor 频率因子
frequency response analyzer 频响分析仪
fresh air 新鲜空气
fresh water rinsing 新鲜水洗
fresh water 淡水
freshen 使清新，使新鲜
fret 磨损[耗]；以格子细工装饰，雕花，侵蚀，磨损，使粗糙
fretage 摩擦腐蚀
fretting corrosion 微动腐蚀
fretwork 浮雕细工，回纹装饰
friability 脆性，易剥落性
friable 易碎的，脆弱的
fricative 摩擦的，由摩擦而生的
friction 摩擦
friction coefficient 摩擦系数
friction element 摩擦元件
friction oxidation 摩擦氧化
frictional 摩擦的，由摩擦而生的
frictional loss 摩擦损失
frictional resistance 摩擦阻力
frictionize 摩擦
frigid 寒冷的，严寒的，冷淡的
frigolabile 不耐寒的
frigorific 致冷的，冰冻的，产生或引起寒冷的
frigostabile 耐寒的
fringe 干涉带，干扰带，条纹，边缘
fringing 边缘通量，散射现象
fritting 熔化，熔结
frogging 互换
frondose 多叶的，叶状的，生叶状体的
front projection 正面投影
front view 正视图

frost 霜，冰冻；结霜，受冻，结霜于
frost boiling 冻胀，冰沸现象
frost crack 冻裂，冻裂隙
frost resistance 抗寒性，耐冻性
frostbound 冰冻的，冻硬的
frosting 结霜，霜状白糖，无光泽面，去光泽；结霜的，磨砂的，消光的
frosted glass 磨砂玻璃
froth 泡沫；起泡沫
froth flotation 浮选
froth promoter 泡沫促进剂
frother 起沫剂
frozen 冻结的
frozen stress 冻结应力
fuel 燃料；供以燃料，加燃料
fuel ash corrosion 燃灰腐蚀
fuel cell 燃料电池
fuel consumption 燃料消耗
fuel cycle 燃料循环
fuel element 燃料元件
fuel ratio 燃料比
fuel reprocessing 燃料再处理
fugacity 逸度，挥发度
fugitiveness 逸散能，不稳定性，不耐久性，易逝性
fugitometer 褪色计，燃料试验计，耐晒牢度试验仪
fulgurant 闪耀的，闪烁的，闪电状的
fulgurate 发光，打闪；发光，用电气烧灼，打闪
full annealing 完全退火
full charge 完全充电
full dip pretreatment system 全浸式前处理
full discharge 完全放电
full dull 全消光
full hardening 完全硬化
full width at half maximum 半高峰宽
fullerence 富勒烯，足球烯，C60
full-scale value 满刻度值
full-wave direct current（FWDC） 全波直流

fulminic 爆发性的
fulvous 黄褐色的，茶色的
fume 烟气
fume cupboard 通风橱
fume extractor 排烟装置，排烟设备
fuming nitric acid 发烟硝酸
fuming sulfuric acid 发烟硫酸
fumigant 熏剂，熏蒸消毒剂
fumous 冒烟的，烟色的
fumy 冒烟的，多烟的，多蒸汽的
function 函数
function generator 函数发生器
functional coating 功能膜
functional group 功能基，官能团
functional ion exchange resin 功能性离子交换树脂
functional materials 功能材料
functional membrane 功能性膜
functionality 官能度
functionalization 官能作用
fundament 基础
fundamental 基本原理；基本的，根本的
fundamental frequency 基频，固有频率
fungicidal 真菌的，杀霉菌的
fungicide 杀真菌剂
fungusized 涂防霉剂的
fungusproof 防菌，抗雾的，防霉的
furan 呋喃
furan carboxylic acid 糠酸
furaldehyde 呋喃甲醛，糠醛
furancarbinol 呋喃甲醇，糠醇
furanose 呋喃糖
furbish 磨光，研磨，擦亮

furcal 叉状的，开衩的，剪刀状的
furcated 叉形的，分叉的
furfurol 糠醛
furl 卷起，折叠，卷起之物；叠，卷收，收拢
furnace 熔炉，火炉
furnace annealer 电热炉退火处理机
furnace cooling 炉冷
furnace lift travel 炉移动行程
furnace lining 炉衬
furring 毛皮，镶边，毛皮衬里，钉板条
furrow 皱纹，犁沟；弄皱
furrowy 多皱纹的，有沟的
fusant 熔化物，熔体
fusation 熔化
fuscous 暗褐色的，深色的
fuse alloy 易熔合金
fuse point 熔点
fuse salt 熔融盐
fused 熔化的，熔凝的，装有保险丝的
fusibility 可熔性，熔度，熔融度
fusible 易熔的，可熔化的
fusible alloy 易熔合金
fusiform 梭形的，纺锭状的，两端渐细的，流线型的
fusing 熔断，熔化，烧断保险丝
fusing agent 熔剂
fusing assistant 助熔剂
fusing point 熔点
fusion 熔融，熔化，熔解
fusion arc welded 熔弧焊
fusion point 熔融点
fusion range 熔融区

G

gage 测量仪表，规格，度量，估计
gage glass 玻璃液位计
gage length 标距，计量长度
gain 增益
gallic acid 鞣酸，没食子酸，五倍子酸
galling 使烦恼的，难堪的，使焦躁的；毛边
gallium 镓
gallon 加仑
galvanic action 电池作用，电偶作用
galvanic cell 原电池，自发电池
galvanic corrosion 电化学腐蚀，接触腐蚀，电偶腐蚀
galvanic couple 电偶对
galvanic deposit 电沉积
galvanic series 电偶序，电位序
galvanize 镀锌，通电，（用电）刺激
galvanized 镀锌的，电镀的
galvanized sheet steel 镀锌钢板
galvanizer 电镀工
galvanizing 热镀锌，热浸镀锌
galvanizing embrittlement 镀锌脆性
galvannealing 镀锌层退火处理，镀锌层扩散处理
galvanochemistry 电化学
galvanometer 检流计，电流计
galvanostatic double pulse method (GDP) 恒电流双脉冲方法
galvanostatic 恒电流的
galvanostatic method 恒电流法
galvanostat 恒电流法，恒电流仪
gamma camera γ射线照相机
gamma equipment γ射线设备
gamma radiography γ射线照相术
gamma ray γ射线
gamma ray radiographic equipment γ射线照相装置
gamma ray source γ射线源

gamma ray source container γ射线源容器
gamma ray unit γ射线机，γ射线装置
gang die 复合模
gap 间隙，缺口，分歧，间隔；造成缝隙，使成缺口
gap grading 不连续粒度，间断级配
gap scanning 间隙扫查，间隙扫描
gas adsorption method 气体吸附法
gas carburizing 气体渗碳法，气体渗碳
gas centrifuge process 气体离心法
gas checking 气体裂纹，气裂
gas chromatography (GC) 气相色谱法，气相色谱分析
gas chromatography-mass spectrometry 气相色谱—质谱分析
gas classification 气体分级法，气体分级
gas constant 气体常数，气体常量
gas crazing 气体裂纹，细裂
gas cutting 气割，氧炔切割
gas hole 气孔
gas mark 焦痕，烧焦，气纹
gas permeability 透气性
gas pickling 气体浸渍法，气体侵蚀
gas porosity 气体孔隙率，疏松度
gas pump 气泵
gas purifier 气体净化器，气体净化设备
gas recombinating efficiency 气体复合效率
gas sensing electrodes 气敏电极
gas shielded arc welding 气体保护焊，气体保护电弧焊
gas vent 气孔，排气孔，通气管，气体出口
gaseous corrosion 气相腐蚀，气相腐蚀
gaseous cyaniding 气体氰化法
gaseous diffusion process 气体扩散法
gaseous state 气态
gasket 垫片
gasket corrosion 垫片间隙腐蚀

gasoline 汽油
gasoline resistance 耐挥发油性，耐汽油性
gasoloid 气溶胶，气胶溶体
gassing 析气，气体处理
gassy 气体的，含气的，气状的
gassy surface 疏松结构
gate 闸门
gatewidth 门脉冲宽度，波门宽度
gauge board 仪表板，样板，模板，规准尺
gauge length 标距
gazette 报，公报
gel 凝胶
gel chromatography 凝胶色谱法，凝胶层析
gel effect 凝胶效应，胶凝效应
gel filtration chromatography (GFC) 凝胶过滤色谱法
gel permeation chromatography (GPC) 凝胶渗透色谱法
gel time 胶化时间，凝胶时间
gelate 胶凝，形成胶体
gelatification 胶凝作用
gelatin 明胶，凝胶
gelatine 胶质，白明胶
gelatineous 胶状的
gelled 有散热片的，装有凸片的
gelled electrolyte 胶体电解液
gelling agent 凝胶剂
gel-type ion exchange resin 凝胶型离子交换树脂
general carbon steel 普通碳素钢
general character 共性，一般特性
general chemistry 普通化学，基础化学
general corrosion 全面普遍腐蚀，普遍腐蚀，均匀腐蚀
general formula 通式
general specification 一般规格，通用规范，一般技术要求
general structure low-alloy steel 普通低合金结构钢

generate 使形成，发生
generator 发电机，发生器
geoline 石油，凡士林
geometric 几何学的，几何的
geometric area 几何面积
geometric unsharpness 几何不清晰度
geometry 几何学［形状，图形，结构］
geotemperature 地温
geotherm 地热
geothermal 地热的，地热能，地温的
ger-bond 热塑性树脂黏合剂
germanate 锗酸盐
germanic （正，含，四价）锗的
germanium 锗
germanite 锗石，亚锗酸盐，二价锗酸盐
germanomolybdate 锗钼酸盐
germanous 二价锗的，含二价锗的
germifuge 杀菌剂
gibbosity 凸圆，突起，软骨病
Gibbs adsorption isotherm 吉布斯吸附等温式
Gibbs free energy 吉布自由能
Gibbs phase rule 吉布斯相律，相律
gibbsite 三水铝矿，水铝矿，三水铝石
giga- 【构词成分】十亿，千兆
gigahertz 千兆赫
gild 镀金，包金，贴金，镀金于，装饰
gilded 镀金的，涂金色的，装饰的
gilding 镀金，虚饰的外观
gilt 镀金，炫目的外表
girth weld 环形焊缝，周围电焊
girth 周长；以带束缚，围绕，包围，围长为
glacial 冰的，冰冷的
glacial acetic acid 冰醋酸，冰乙酸
gland （密封）压盖，衬垫
glandless 无垫料的，无密封垫的
glass electrode 玻璃电极
glass fiber 玻璃纤维
glass fiber reinforced plastics 玻璃钢，玻璃纤维增强塑料

glass fiber separator 玻璃纤维隔板
glass funnel long stem 玻璃漏斗
glass rod 玻璃棒
glass state 玻璃态
glass transition point 玻璃转化点
glass-ceramic 玻璃陶瓷
glassy 玻璃质,像玻璃的
glassy carbon electrode 玻碳电极
glazing 光滑,上光,上釉
glazy 上过釉的
globose 球状的,球形的
globosity 球形,球状
globular 球状的,由小球形成的
globular cementite 球状炭化铁
globularity 球状,球形
globurizing 球状化,球化
gloea 黏质,胶
glomerate 集合的,密集成簇的,团聚的
glomeration 球状体,成球状,球状化,聚合,黏结,团聚（之物）
glomic 球的
gloom 阴暗,使黑暗；
gloomy 黑暗的
gloppy 糊状物质,黏糊糊的
gloss 光泽；使光彩
gloss finish 光泽精饰,光泽整理
gloss meter 光泽计,光泽度计
glossiness 光洁度,光泽度,有光泽
glossing （平滑表面上）光泽,光亮；
glossreducing agent 消光剂
glossy 有光泽的,光亮的
glove valve 球阀
glow discharge 辉光放电
glow discharge heat treatment 辉光放电的热处理
glow discharge mass spectrometry (GDMS) 辉光放电质谱
glow discharge nitriding 辉光放电氮化,离子氮化
glow discharge optical emission spectrometry 辉光放电发射光谱
glow discharge spectrometry (GDS) 辉光放电光谱学
glucan 葡聚糖
gluconic 葡萄糖的
gluconic acid 葡糖酸
glucose 葡萄糖
glucoside 葡萄糖苷,配糖体
gluside 糖精
glutaraldehyde 戊二醛
glutaric acid 戊二酸,谷酸
glutinosity 黏性,黏着性
glyceride 甘油酯,甘油化物
glycerinated 含甘油的,甘油制的
glycerine 甘油,丙三醇
glycerol 甘油,丙三醇
glycerol monolaurate 甘油单月桂酸酯
glycerol monostearate 单硬脂酸甘油酯
glycerol stearate 硬脂酸甘油酯
glycidic acid 环氧丙酸,缩水甘油酸
glycidyl 环氧丙基,缩水甘油基
glycine 甘氨酸,糖胶,氨基乙酸
glycine betaine 甜菜碱
glycoside 糖苷
glycosyl- 【构词成分】糖基
glyoxal 乙二醛
glyoxaline 咪唑
glyoxylic acid 乙醛酸,水合乙醛酸
glyphographic 电雕版术的
gofer 皱纹；使起皱褶［皱纹］
goggles 护目镜
gold 金
gold decoration 金修饰
goldplate 镀金
goldplated 镀金的
gole 溢水道,溢流堰
gouge 沟,圆凿,以圆凿刨；用半圆凿子挖
gouging 刨削槽
governor 调节器,控制器
gracile 细长的,薄的

grad cylinder 量筒
gradation 级配
gradient 梯度
grading 分类，分阶段，定等级
grading analysis 粒度分析
graduated 分等级的，分度的，刻度的
graduated cylinder 量筒
graduated pipette 刻度吸管
graniferous 粒状的，有颗粒的
graft 移植
graft copolymer 接枝共聚物
grain 晶粒，颗粒
grain boundary 晶界
grain boundary attack 晶界侵蚀，粒界腐蚀，晶界腐蚀
grain boundary crack 晶界裂纹，晶粒间界开裂，晶粒间界裂缝
grain boundary reaction 晶界反应
grain coarsening 晶粒粗化
grain coarsening temperature 晶粒化温度
grain fineness number 晶粒度，细度
grain growth 晶粒长大，晶粒生长
grain refinement 晶粒细化
grain size 结晶粒度，晶粒尺寸
grain size strengthening 晶界强化
graininess 粒状，多粒，木纹状的，微粒状态
graining out 析皂
graining point 析皂点
gram 克
granodising 磷酸盐处理（锌及镀锌制品油漆前的处理）
granular 颗粒状的，粒状的
granular activated carbon 颗粒状活性炭，粒状活性炭
granular activated carbon adsorption tank 颗粒活性炭吸附池
granulation 粒状化，造粒，使成粒状
grape sugar 葡萄糖
graphic 图解的，用图表示的，用文字表示的，形象的，生动的

graphite 石墨；用石墨涂（或搀入等）
graphite electrode 石墨电极
graphite flake precipitation 片状石墨析出
graphitic 石墨的
graphitic corrosion 石墨化腐蚀，碳化腐蚀
graphitic oxide 氧化石墨，石墨氧化物
graphitization 石墨化
graphitized 石墨化的
graphitizing 石墨化，石墨化作用
graphitizing treatment 石墨化退火，石墨化处理
grate plane 晶格面
grating 光栅，摩擦，摩擦声，格子；摩擦的
grating constant 晶格［点阵，光栅，格栅］常数
grating spectrometer 光栅光谱仪，光栅分光计
gravelometer 石击崩裂试验法
gravimetric （测定）重量的，重量分析的
gravimetric analysis 重量分析法
gravimetric method 重量法
gravimetric titration 重量滴定
gravimetry 重量分析
gravity 重力，比重，严重（性）
gravity casting 重力铸造
gravity filter 重力式过滤器
gravity segregation 垂力偏析，密度偏析
gravity sintering 重力烧结
gravity spray gun 重力式喷枪
gray cast iron 灰口铸铁
grazing angle 掠射角，切线角，入射余角
grazing incidence 掠入射，切线入射，临界入射
grease 油脂；涂脂于
grease and oil 油脂
grease pit 污斑
greasing 润滑，涂油，起油腻；润滑的，油脂的
greasy 油腻的，含脂肪多的

greenish 呈绿色的，浅绿色的
green-weight 湿重
gregaloid 集合样的，群状的，簇聚的
grey 灰色；灰色[白]的；使变成灰色
grey cast iron 灰铸铁
grey scale 灰度
grid 网格，格子，栅格
grid plate 格网板，（蓄电池）涂浆极板，铅板
grill 烧，烤
grind 摩擦，磨碎
grind gauge 细度刮板
grindability 打磨性，易磨性
grinder 粉碎机，磨床，砂轮机，研磨机
grinding 磨光，研磨
grinding aids 研磨助剂
grinding defect 磨痕
grinding disc 研磨盘，砂轮
grinding machine 研磨机，磨床
grinding material 磨料
grinding scratch 锉刀纹
grinding tool 磨削工具
grinding wheel 磨轮，砂轮片，砂轮
grindstone 磨石
grip 紧握，柄，支配；紧握，夹紧
gripper 夹具
griseous 珍珠灰的，浅灰色的
grit 砂砾，研磨[石英]砂，磨粒[料]；研磨，在……上铺砂砾
grit blasting 喷钢砂，喷丸，喷砂处理
grit size 粒度，粒度号，砂粒尺寸，磨料粒度
gritty 多沙的，砂砾质的
grizzley 带灰色的

groove 凹槽，压线；开槽于，形成沟槽
grooving 挖槽的
grooving corrosion 沟蚀，沟纹腐蚀
gross error 过失误差
ground electrode 接地电极，地线
ground state 基态
group 族
growth 生长，胀大
growth mechanism 生长机制，生长机理
growth orientation 生长取向
growth pattern 生长模型，生长模式
growth rate 生长速率，增长率
guide 指[引，制，向，教]导，支配，控制，导向，滑槽
guide coat 二道浆
guiding 导向，控制，定向；引导的，控制的，制导的，导向的
gum 树胶；用胶粘，涂以树胶，使……有黏性
gum arabic 阿拉伯胶
gummiferous 含胶的
gumminess 黏性，树胶质，胶黏性，黏着
gumming 树胶分泌，涂胶
gumming test 结胶试验，胶黏试验
gummosity 黏着性，胶黏性
gummous 似树胶的，有黏性的
gun color 枪色
gun distance 喷枪距离
guttate 有斑点的，滴状的
gutter 排水沟，槽；流，形成沟，开沟于……
guttiform 滴状的，点滴形的
gypsum 石膏；用石膏处理
gyroscopic 回转的，回转仪的，陀螺的

H

hacksaw 钢锯；用钢锯锯断
hair cracking 鬈状裂纹，发裂，发纹
hair line 毛纹，发线，细微裂纹
hair seam 毛状叠纹
haircrack 发裂，发状裂缝，毛细裂纹
hair-line finish 发纹处理
half cell potential 半电池电势
half dip pretreatment system 半浸式前处理
half extinction 半消光
half life 半衰期
half step potential 半波电势
half wave width 半波宽度
half-cell 半电池
half-reaction 半反应
half-value method 半波高度法
half-wave current 半波电流
halide 卤化物，卤素的
halo- 【构词成分】卤，含卤素的
haloalkylation 卤烷基化
halochromic 加酸显色的，卤色化作用的
haloform 卤仿，三卤甲烷
haloform reaction 卤仿反应
halogen 卤素
halogen hydride 卤化氢
halogenate 卤化
halogenated hydrocarbon 卤代烃
halogenation 卤化，加卤作用
halogenic 卤素的，生盐的
halogenide 卤化物
halogenous 含卤的
hammermill 锤磨机
hamular 钩状的
hamulate 弯曲的，顶端弯折的，有钩的
hamulus 钩，钩状突起
hand charging 手工加料
hand electrostatic spray equipment 手提式静电喷涂装置

hand electrostatic spray gun 手提式静电喷枪
hand forging 手锻
hand molding 手造模，手工造型
handbook 手册，指南
handily 方便地，敏捷地，灵巧地
handiness 轻便，灵巧，操纵方便，易操纵性
handlance 手喷枪，手压泵，喷枪
handleability 可操作性，易操作性，操纵性能
hanger 吊架，挂架，洗片架
hanging 悬挂；悬挂着的
hanging mercury drop electrode（HMDE） 悬汞滴电极
hard anodic oxide coating 硬质阳极氧化膜
hard anodizing 硬阳极氧化处理法，硬质阳极氧化
hard case 表面渗碳硬化
hard chrome plating 镀硬铬
hard cold work 硬级冷加工
hard drawing 硬拉，硬抽
hard magnetic material 硬磁材料
hard metal 硬质合金
hard rubber container 硬橡胶槽
hard solder 硬焊料，钎焊
hard water 硬水
hard wrought 冷加工的，冷锻的
hard zone crack 熔接区裂纹
hardenability 淬透性，淬硬性
hardenability band 淬透性带
hardenability curve 硬化性曲线，淬透性曲线
hardenable 可硬化的
hardened structure 硬化组织，淬火组织
hardener 硬化剂
hardening and tempering 调质
hardening capacity 淬硬性

hardening furnace 淬火炉
hardening penetration 硬化深度
hardening temperature 硬化温度
hardening thickness 硬化厚度
harder 固化剂
hardness 硬度
hardness ageing 加工时效，硬化时效
hardness factor 硬度因数
hardness gage 硬度计
hardness test of anodic oxide coating 阳极氧化膜硬度试验
hardness test 硬度试验
haring cell 哈林槽
harmonics 谐频，谐波
harmonic analysis 谐波分析
harness 利用
harsh 严厉的，粗糙的
harsh working 难以加工的
hastelloy 耐蚀耐热镍基合金
hatching 剖面线，阴影线
haze 雾状
hazy 模糊的，有雾的
head tank 高位（水）箱
health hazards 健康危害
heap leaching 堆集浸滤，堆集浸洗
hearth 炉床，炉底
heat affected area 热影响区
heat affected zone (HAZ) 热影响区，高热影响区（焊接件的）
heat balance 热平衡，热量衡算
heat checking 热裂，热裂纹
heat conduction 热传导
heat conductivity 热导率
heat cycle test 热循环试验
heat dissipation 热传，散热
heat exchanger 热交换器
heat insulating material 绝热材料，保温材料
heat loss 热（量）损失，热损耗
heat of dissolution 溶解热

heat of evaporation 蒸发热
heat of formation 生成热
heat of friction 摩擦热
heat of fusion 熔化热
heat of hydration 水化热，水合热
heat of ionization 电离热
heat of neutralization 中和热
heat of reaction 反应热
heat of solidification 凝固热
heat of sublimation 升华热
heat preserving furnaces 保温炉
heat pump 热泵，蒸汽泵
heat recovery 热回收
heat resistance 耐热性
heat resistant alloy 耐热合金
heat resistant steel 耐热钢
heat resisting coating 耐热涂料
heat sink 散热件［片］
heat spotting 局部热处理
heat stabe maskingtape 耐热遮盖用胶带
heat tinting 热蚀法，热染，热着色，回火颜色
heat transfer coefficient 热传递系数
heat transfer printing 热转移印染
heat transfer 热传输，热传递
heat transmission 热传递，热传导
heat treating film 热处理膜
heat treatment 热处理
heat treatment furnace 热处理炉
heat treatment in fluidized beds 流态床热处理
heat treatment in protective gases 保护气氛热处理
heat value 热值，卡值
heat-proof paint 耐热漆
heated 热的
heater cooler 加热器冷却装置
heating 加热；加热的，供热的
heating element 发热元件，加热元件
heating furnace 加热炉
heating oil separate 加热油水分离装置

heating treatment furnace 热处理炉
heating zone 加热区加热带
heavy media separation 重介质分离，重介选
helical etching system 螺旋型蚀刻系统
heliotropin 3,4-二氧亚甲基苯甲醛，洋茉莉醛，胡椒醛
helium 氦
hematite 赤铁矿
hemiacetal 半缩醛
hemicolloid 半胶体
hemihedry 半面形，半对称
hemi-homolysis cleavage 半均裂
hemispherical sector analyser (HSA) 半球型能量分析器
hemitrisulfide 三硫化二物
hemming 褶边，卷边
hendec(a)- 【构词成分】十一
hendecagon 十一角形，十一边形
hendecahedron 十一面体
Henry's constant 亨利常数
Henry's law 亨利定律
Herber hardness tester 哈柏特硬度试验机
hermetic 密封的，与外界隔绝的，不透气的
hermetic package 气密封装
hermetic seal 密封，气密封口
hermetically tight seal 气密密封，密封，密封装置
heter(o)- 【构词成分】杂，异，不同，不均一
hetero epitaxial 异质外延的
heteroacid 杂酸
heterocrystal 异质晶体，异种晶体
heterocycle 杂环
heterocyclic 杂环的，不同环式的
heterocyclic compound 杂环化合物
heterodisperse 非均相分散，杂分散
heteroepitaxy 异质外延，异质磊晶，异质外延法
hetero geneous membrane electrodes 非均相膜电极
heterogeneity 非均质，不均匀性
heterogeneous 非均相，多相的，不同种类的，异类的
heterogeneous catalysis 非均相催化剂
heterogeneous equilibrium of ions 多相离子平衡
heterogeneous hydrogenation 多相氢化
heterogeneous nucleation 非均质成核
heterogeneous reaction 异相反应
heterogeneous structure 非均质组织
heterolysis 异裂，异种溶解，外力溶解
heteromorphic 异型的，多晶型的，多晶的
heteromorphism 异态性，多晶现象，异形性
heteromorphous 多晶的，异态的
heterophase 非均匀相结构，非匀称相结构
heteropolyacids 杂多酸
heteroside 糖杂体，葡萄糖甙
heterostructure 异质结构
hex(a)- 【构词成分】六，已
hexabasic 六（碱）价的，六代的，六元的
hexadeca- 【构词成分】十六
hexagon 六角形，六边形；成六角的，成六边的
hexagonal 六角形的，六角的
hexagonal close-packed (HCP) 密排六方结构
hexagonal system 六方晶系
hexahydrate 六水合物
hexamer 六聚物，六聚体
hexametaphosphate 六偏磷酸盐
hexamethylenetetramine 六亚甲基四胺，乌洛托品
hexamine 六胺，乌洛托品
hexanal 己醛，正己醛，乙醛
hexandioic acid 己二酸
hexanitro- 【构词成分】六硝基

hexaplanar 六角晶系，平面六角晶，六角平面的
hexapolythionate 连六硫酸盐
hexavalent chromium 六价铬
hexoxide 六氧化物
hiatus 裂缝，空隙，脱漏部分
hierarchical 体系的，分层的，分级的，等级的
hiding chart 遮盖力试验纸
hiding failure 遮盖不良
hiding power 遮盖力
high alloy steel 高合金钢
high boiling point solvent 高沸点溶剂
high carbon steel 高碳钢
high density polyethylene（HDPE） 高密度聚乙烯
high energy heat treatment 高能束热处理
high energy X-ray 高能 X 射线
high frequency generator 高频振荡器，高频发生器
high frequency heating 高频加热法
high frequency induction heating method 高频感应加热法
high frequency titration 高频滴定
high impact polystyrene（HIPS） 高冲击聚苯乙烯，耐冲击聚苯乙烯
high orded pyrolytic graphite 高序热解石墨
high performance capillary electroporesis 高效毛细管电泳法
high performance liquid chromatography 高效液相色谱法
high performance liquid chromatography-mass spectrometry 高效液相色谱－质谱联用
high polymer 高聚物
high pressure jet spray 高压喷射喷雾机
high pressure method 高压法
high pressure operation 高压操作
high pressure steam 高压蒸汽
high rate discharge 高效率放电，大电流放电
high rate discharge characteristic 高效率放电特性
high resolutionelectron energy loss spectroscopy 高分辨电子能量损失光谱
high silicon cast iron 高硅铸铁
high solid lacquer 高固体分漆
high solid paint 高固体分型涂料
high speed electrodeposition 高速电镀
high speed steel 高速钢
high strength low-alloy steel 高强度低合金钢
high strength steel 高强度钢
high sulfur nickel 高硫镍
high-temperature carburizing 高温渗碳
high temperature corrosion 高温腐蚀
high temperature manganese phosphating 高温锰系磷化
high-temperature oxidation 高温氧化
high temperature superconducting material 高温超导材料
high temperature tempering 高温回火
high tensile coldrolled steel sheet 冷轧高强度钢板
high vacuum 高真空
high voltage electron microscopy 高压电子显微镜学
high-energy ion-scattering spectrometry（HEISS） 高能离子散射谱
high-frequency furnace 高频炉
highly ordered pyrolytic graphite 高度有序热解石墨，高序热解石墨
high-quality carbon steel 优质碳素钢
high-speed transmission 高速传递，高速传输
hilsonite 天然硬沥青
hindered amine light stabilizer 受阻胺类光稳定剂
histogram 柱状图
hizex 高密度聚乙烯
hoevellite 钾盐

holding device 夹具，夹持装置
holding furnace 保温炉
hole 空穴
holey 多洞的，有穴的
hollow molding 中空（吹出）成形
holo- 【构词成分】完全，整体
holocrystalline 全晶质的，全晶的
hologram 全息图，全息摄影，全息照相
holographic 全息的
holography 全息照相术
holohedral 全对称晶型的，全面的
holohedrism 全对称性
holohedron 全面体
holomorphic 正则的，全形的
holonomy 完整
holoside 多糖
holystone 磨石，用磨石墨
homaxial 等轴的
homedric 等平面的
homenergic 等能量的
homeo- 【构词成分】相同，相等，类似
homeomorphism 同胚，异质同晶
homeomorphy 异质同晶
homeomorphous 同形（态）的
homeosmoticity 恒渗（透压）性
homeostasis 体内平衡，自动（调节）动态平衡，稳衡
homeostat 同态调节器
homeostatic 稳态的
homo- 【构词成分】（相）同，相（类）似，共同，同质［型，一］，均匀，高，升，连合
homoallylic 高烯丙基
homoallylic alcohol 高烯丙醇
homoazeotrope 均相共沸混合物
homodisperse 均相分散，均匀分散
homoenergetic 均［同，高］能的
homoeomorphic 异质同晶的，同形态的
homoepitaxy 均相外延
homogen 均质，均质合金
homogeneity 均质，同质

homogenization 均化，匀化
homogeneous 均匀的，同质的，同类的
homogeneous carburizing 均质渗碳
homogeneous catalysis 均相催化剂
homogeneous deformation 均匀变形
homogeneous material 均质材料
homogeneous nucleation 均态成核
homogeneous structure 均质组织
homogeneous hydrogenation 均相氢化
homogeneous membrane electrodes 均相膜电极
homogeneous reaction 均相反应
homogeneously 同类地，相似地，均一地，均匀地
homogenization 均质化，均质作用
homogenizer 均化剂，均质器，高速搅拌器
homogenizing 均质化，均质［扩散］退火
homogenizing anneal 均质化退火
homo-ion 同离子
homoiosmotic 恒渗压的，恒渗透压的，等渗性的
homoiothermal 恒温的
homoiothermic 调温的
homolateral 同侧的
homolog 同系物
homologization 均裂作用
homologous 相应的，同源的，类似的，一致的
homologue 同系［对应，相似］物
homology 同源性，同调，同系物
homolysis 均裂，同种溶解，同族溶解
homolytic 均裂的
homolytic cleavage 均裂
homomorphism 同态，同晶型，同形
homopolymer 均聚物
homo-treatment 均匀热处理
hondrometer 粒度计，微粒特性测定计
hone 磨刀石，金属表面磨损
honeycomb 蜂窝［器，结构，格栅，状砂眼］；使成蜂巢状

honeycomb cracks 龟裂，网状裂隙
hood 通风橱，排气罩
hook 挂钩，吊钩；钩住
hook lock 钩锁
hopper 加料斗，漏斗
hot acid circulation and filtration system 高温酸液循环过滤系统
horizontal 水平的
horizontal line 水平线，横线
hot air circulating equipment 热风循环装置
hot air convection oven 对流炉
hot air drying convection drying 热风干燥，对流干燥
hot bath quenching 热浴淬火
hot bent test 热变试验
hot corrosion 热腐蚀
hot cracking 热裂，热裂纹
hot deformation 热变形
hot dipping 热浸镀
hot drawing 热拉法
hot extrusion 热挤压
hot forging 热锻造
hot forming 热成形
hot hardness 高温硬度
hot heading 热锻粗
hot hubbing 热模压
hot lime process 热石灰法
hot mark 热斑
hot melting 热熔
hot peening 热珠击法，热锤击
hot quenching 高温淬火
hot rolled steel sheet 热轧钢板
hot spraying 热喷涂
hot stamping foil 烫金膜
hot stamping 烫金，烫印
hot stretching 热伸展，热拉伸
hot tinting 热着色法
hot water prerinsing 预热水洗
hot water resistance test 耐温水试验
hot working 热加工

hotrolled steel sheet 热轧钢板
hpdrocarbon 碳氢化合物
hue 色相，色调，饱和度
Hull cell 赫尔槽
humidity 湿度
humidity resistance test 耐湿试验
hybrid 混合的，杂种的
hybrid composite 混杂复合材料
hybrid orbital 杂化轨道
hybridization 杂化
hybridization affect 杂化影响
hydration 水合，水合作用
hydraulic 液压的，水力的，水力学的
hydraulic classification 水力分层［级］
hydraulic power tool 液压工具
hydraulic component 液压元件
hydrazine 肼，联氨
hydride 氢化物
hydride descaling 氢化物去锈法
hydroacylation 加氢酰化
hydro-blast 高压水清砂机
hydroboration 硼氢化反应，硼氢化作用
hydrocarboxylation 氢羧基化
hydrochloric acid 盐酸
hydrocracking 加氢裂化，氢化裂解
hydrodynamic method 流体动力学方法
hydroformylation 加氢甲酰基化
hydrogen 氢气，氢
hydrogen attack 氢腐蚀
hydrogen blistering 氢鼓泡
hydrogen bond 氢键
hydrogen damage 氢损伤
hydrogen electrode 氢电极
hydrogen embrittlement 氢脆
hydrogen evolution 析氢
hydrogen flame ionization detector 氢焰离子化检测器，氢火焰离子化分析仪
hydrogen induced cracking 氢致开裂
hydrogen induced delay cracking 氢致延迟断裂
hydrogen overvoltage 氢过电位

hydrogen peroxide 双氧水，过氧化氢
hydrogen relief annealing 预防白点退火，脱氢退火
hydrogen scale 氢标
hydrogenolysis 氢解
hydrolysed 水解的
hydrolysis 水解作用
hydrolyzate 水解产物
hydrometallation 氢金属化
hydrometallurgy 水冶金，湿法冶金
hydrometer 比重计，液体比重计
hydronium ion 水合氢离子
hydroperoxide 氢过氧化物
hydrophile 亲水物
hydrophilic 亲水的
hydrophilic emulsifier 亲水性乳化剂
hydrophilic group 亲水基
hydrophilic remover 亲水性洗净剂
hydrophilic-lipophilic balance 亲水-亲油平衡值
hydrophilic-lipophilic ratio 亲水-亲油比
hydrophily 亲水性
hydrophobic 疏水的，憎水的
hydrophobic agent 疏水剂
hydrophobic group 亲油基，疏水基
hydrophoby 疏水性
hydroplastic forming 含水塑形成形
hydroscope 验湿器，水气计，湿度计
hydroscopic 吸湿的，吸水的
hydrosol 水溶胶，胶质溶液
hydrosoluble 水溶性的，可溶于水的
hydrosolvent 水溶剂
hydrostatic 流体静力学的，静水力学的
hydrostatic extrusion 静力液挤压
hydrostatic pressing 液均压法
hydrostatic tension 均张力，静液压应力，静液压张力
hydrosulphate 硫酸氢盐
hydrosulphite 亚硫酸氢盐
hydrotrope 水溶物
hydrotaxis 趋水性，向水性

hydro-thermal sealing 水合封孔处理
hydrous 水合的，含（结晶）水的
hydroxide 氢氧化物
hydroxyalkylation 羟烷基化
hydroxyethyl 羟乙基纤维素
hydroxyl 羟基
hydroxylation 羟基化
hydroxymethylation 羟甲基化作用
hygrometer 湿度计
hylotropy 恒沸性，恒熔性
hyperacidity 酸过多
hyperconcentration 超浓缩
hypereutectic 过共晶；过共晶的，过低熔的
hypereutectic alloy 过共晶合金
hyperchromic effect 增色效应，浓色效应
hypereutectoid 过共析体，高共析质
hypereutectoid alloy 过共析合金
hyperfiltrate 超滤
hyperfine 超精细的
hypochloric acid 次氯酸
hypochlorite 次氯酸盐
hypochromic 着色不足的
hypochromic shift 紫移
hypochromic effect 减色效应，淡色效应
hypoeutectic 低级低共熔体；亚共晶的
hypoeutective alloy 亚共晶合金
hypoeutectoid 低易熔质；亚共析的
hypoeutectoid alloy 亚共析合金
hypoeutectoid steel 亚共析钢
hypoiodite 次碘酸盐
hypoosmotic 低张溶液的，低渗的
hypophosphate 连二磷酸盐，次磷酸盐
hypophosphite 次磷酸盐
hyposaline 低盐度
hyposteel 亚共析钢
hyposulphite 硫代硫酸盐，连二亚硫酸盐
hypothermal 低温的，降温的
hypothermy 低温，降温
hypotonicity 低渗，低张性，张力减退
hypovanadate 次钒酸盐

hypovanadic oxide 二氧化钒
hypsochrome 浅色团，浅色基，紫移色团
hypsochromic 向蓝移（的）

hypsochromic shift 短移
hysteresis 磁滞，滞后

I

ice machine 制冰机，冷冻机
ichnography 平面图，平面图法
icon 图标［像］，插图
icos(a)- 【构词成分】二十
icosagon 二十边形
icosahedral 二十面体的
icosahedron 二十面体
icos(a)- 【构词成分】二十
icy 冰冷的，冷淡的，结满冰的
ideal depolarized electrode 理想去极化电极
ideal nonpolarizable electrode 理想非极化电极
ideal polarized electrode (IPE) 理想极化电极
idealine 糊状黏结剂
ideally 理想的
identical 同一的，恒定的，相同的
identification mark 识别标志
identification 鉴定，辨别，验明
identifier 鉴别器
identify pulse 识别脉冲
idio- 【构词成分】自［身，发］，原有，专有，同，特殊
idiopathetic 自发的，特发的
idiostatic 同电位的，等位差的，同势差的
idiostatic method 同位差法，等电位法，同势差连接法
idiosyncrasy 特质，气质，风格
idiosyncratic 特质的，特殊的，异质的
idiovariation 自发性变异，个体突变，突变
idler 惰轮，中间齿轮
igloss 灼减，减量
igelite 聚氯乙烯塑料
igneous 火的，火成的，似火的
ignescent 发火物；敲击发出火花的
ignitability 可燃性，易燃性，可点燃性
ignitable 可着火的，可燃性的

ignite 点燃，使燃烧
ignitibility 可燃性，易燃性
ignitible 可着火的，可燃的
ignition 灼烧
ignitron 引燃管，点火管
ignitron pulse 触发脉冲，启动脉冲
illegitimate 非法的，不合理的
illimitable 无限的，无边际的
illium 镍铬合金
illuminable 可照明的，可被照明的
illuminance 照（明）度
illuminant 发光体，光源；照明的，发光的
illuminate 阐明，说明，照亮，用灯装饰；照亮
illumination 照明［度］，启发，灯饰（需用复数）
illuminating lamp 照明灯泡
illuminometer 照度计
illusion 幻觉，错觉
illusive 错觉的，幻影的
illustrate 举例，阐明，举例说明，图解
illustration 说明，插图，例证，图解，具体说明
illustrative 说明的，作例证的，解说的
illustrative diagram 原理图，直观图
illustrious 著名的，杰出的，辉煌的，有光泽的，明亮的
illuvial horizon 淀积层，沉积层，聚积层
illuviate 经受淀积作用
illuviation 淀积（作用），淋积（作用）
illuvium 淋积层
ilminite 铝电解研磨法
image amplifier 图像放大器，影像
image analysis system 图像分析系统
image magnification 图像放大
image clarity 影响清晰度
image contrast 图像对比度，影像对比

image converter 影像转换器
image definition 图像清晰度，像清晰度
image depth profile 成像深度剖析
image enhancement 图像增强
imaginary 虚构的，假想的，想象的，虚数的
imaginary component 虚分量，虚数部分，虚部
imaging line scanner 图像线扫描器
imbalance 不平衡，失衡，失调
imbibe 吸收［入，取，液］，浸透
imbibing 吸液［吸收］作用
imbibition 吸入，吸取
imbricate 重叠成瓦状的；边缘重叠成瓦状
imbue 灌输，使感染，使渗透
Imhoff tank 双层沉淀池
imidazole 咪唑，异吡唑
imidazoline 咪唑啉，间二氮杂环戊烯
imide 二酰亚胺
imido 亚氨，酰亚胺的，亚氨基
imine 亚胺
imino- 【构词成分】亚氨基
iminourea 亚氨基脲，亚胺脲，亚氨脲，胍
imitable 可模仿的
imitate 模仿，仿效，仿造，仿制
imitation 仿，仿造，仿制品；人造的，仿制的
imitative 模仿的，仿制的
immaculate 完美的，洁净的，无瑕疵的
immalleable 无展性的，不可锻的
immanence 内在，内在性，固有
immanent 内在的，固有的
immaterial 非物质的，无形的，非实质的
immature 不成熟的，未成熟的，粗糙的
immeasurable 无限的，不可计量的，不能测量的
immense 巨大的，广大的
immensity 巨大，无限，广大

immerse 沉浸，使陷入
immerseable 可浸入的
immersed 浸入的，专注的
immersible 可浸入的，可没入的
immersion 浸没，浸渍
immersion plating 浸镀，浸渍涂敷
immersion probe 液浸探头，水浸探头
immersion rinse 浸没清洗，浸液清洗
immersion test 浸泡实验，浸渍试验
immersion wet etching system 浸渍式蚀刻系统
immiscibility 不混溶性，不混合性
immiscible 不融和的，不能混合的，非互溶的，非搅拌的
immiscible metal 难混溶金属
immission 注入，注射
immittance 导抗，阻纳
immittance chart 阻抗导纳图
immix 掺和，混合，卷入
immixable 不能混合的
immobile 固定的，稳定的，不变的
immobility 不动（性），固定（性）
immobilization 使停止流通，固定，降低流动性，缩小迁移率
immovability 不变，不动
immovable 不动的，固定的，不改变的
immutability 不变，永恒性，不变性
immutable 不变的，不可变的，不能变的
impact 冲击，碰撞，压紧
impact cleaning 抛丸喷射清理，抛丸清理
impact energy 冲击能
impact extrusion 冲击挤压加工
impact fluorescence 轰击荧光
impact resistance 耐冲击性，抗冲强度
impact strength 冲击强度
impact stress 冲击应力
impact tester 冲击试验仪
impacted 压紧的，结实的，嵌入的
impacter 冲击器
impaction 压紧，装紧，嵌入
impairment 损伤，损害

imparity 不平等，不同，不配合
impartible 不可分的，不能分割的
impaste 使成糊状，用糨糊封，涂厚色彩于
impasto 厚涂的颜料，厚涂颜料的绘画法
impatency 闭阻，阻塞，关闭，不通
impatent 闭阻的，非开放性的
impeccable 无瑕疵的，没有缺点的
impedance 阻抗，全电阻
impedance coupling 阻抗耦合
impedance matching 感应淬火，阻抗匹配
impedance plane diagram 阻抗平面图
impedance technique 阻抗技术
impedance test 阻抗试验
impedance transducer 阻抗传感器
impede 阻［妨］碍，阻止
impediment 妨碍，阻止
impedimeter 阻抗计
impedimetry 阻抗滴定法
impeller 叶轮，泵轮，推进器
impeller head 抛丸器［头］
impeller passage 叶片间距
impenetrability 不可入性，不渗（透）性
impenetrable 不能通过的，不能穿过（透）的
imperceptible 感觉不到的，极细微的
imperfection 不完整性，缺陷
imperforate 无孔的
imperious 不透水的
impermeability 不渗透性，不透过性，不能渗透的性质或状态
impermeable 不渗透性的，（对水等）不能渗透的，防水的，密封的
impervious 不能渗透的，不透水的，抗渗的，不透性的
impinge 撞击，侵犯
impingement 冲击，影响，侵犯
impingement angle 入射角，碰撞角
impingement attack 冲击腐蚀，撞击腐蚀，浸［滴，腐］蚀
impingement corrosion 冲蚀

implant 植入物，植入管；种植，灌输，嵌入
implantation 插入，安放，植入
implemental 器具的，有帮助的，可作为工具的
implosion 内爆［裂］，从外向内的压力作用，挤压，冲挤，压碎
imponderability 无重量性，无法测知的性质，失重
imponderable 无法衡量的事物；无重量的，无法计算的
imporosity 无孔性，不透气性
imporous 无孔隙的
import licence position 进口许可证
impound 蓄水
impoverishment 固态合金损失，贫化，损耗
imprecise 不精确的，不严密的，不确切的
imprecision 不精确，不严密
impregnability 坚固性，浸透本领［性能］
impregnable 可浸透的，充满的，坚固的
impregnant 溶剂，浸渍剂
impregnate 浸透，浸渍
impregnated cathode 浸渍阴极，浸渍式阴极
impregnating compound 浸渍剂，浸渍化合物，防腐剂
impregnation 浸胶，浸渍，饱和，充满
impress 压痕，特征，痕迹；盖印，划［压］痕
impression 型腔，模槽，痕迹，底［漆］层，印痕
impression technic 印模术
impression tray 印模盘
impulsator 脉冲发生器
impulse 冲动，搏动，脉冲
impulse eddy current testing 脉冲涡流检测
impulse generator 脉冲发生器
impulse oscilloscope 脉冲示波器
impulse recorder 脉冲记录器

impulse scaler 脉冲计数器
impulse timer 脉冲定时器，脉冲计时器，脉冲计数器
impulse transmitting tube 脉冲发射管
impulser 脉冲发生器，脉冲传感器
impulsing 发出脉冲，脉冲激励
impulsing relay 冲击继电器，脉冲断电器
impulsion 脉冲，冲动［击］
impulsive 冲动的，脉冲的，冲击的
impunctate 无细孔，非点状的
impure 不纯的，污染的，掺杂的，杂质的，混合的
impurity 杂质，混杂物
in parallel 并联，并行
in phase 同相的
in quadrature 正交
in situ 现场，原位，限于原来部位
in stock 库存
in toto 全，整体
in vacuo 在真空中
inaction 无作用，不活动
inaction period 无作用期间，钝化周期，无作用时期
inactive 不活跃的，不活动的，迟钝的
inactive anode 钝化阳极
inbeing 本质，内在的事物，固有（性）
incandesce 白热化，使白热化，炽热化，灼烧
incandescent 辉耀的，炽热的，发白热光的
incandescent lamp 白炽灯
incendive 引火的，易燃烧的
incessancy 持续不断，频繁性
incessant 不断的，不停的，连续的
inch 英寸
incidence 入射，入射角，发生率，影响范围
incident illumination 入射光
incidental 偶发的，非主要的，附带的
incinerate 焚化，烧成灰，把……烧成灰，烧弃

incineration 焚化，烧成灰
incinerator 焚化炉，焚烧装置
incisal 切开的，切割的
incise 切，切割，雕刻
incision 雕刻，切割［开，口］
incisive 敏锐的，锋利的
inclination 倾斜，斜度
inclosed 密闭的，密封式的
inclosure 附件，围绕，罩，壳，包裹体
include 包括，计入
inclusion 包含，包埋，杂质，夹杂物
indicated light 指示灯
indicator 指示器，显示器，指示剂，指针
indicating lamp 指示灯
incoagulability 不凝结性，不能凝固
incoagulable 不能凝固的，不凝的
incoercible 不能压缩成液态的
incoherence 不连贯，不黏结性，无内聚性，松散
incoherent 不连贯的，无黏性的，不胶结的，松散的，无内聚的
incoherentness 不相干性，无黏结性，无内聚性
incohesive 无凝聚力的，无黏聚力的
incombustibility 不燃性，不可燃性，不可燃烧性
incombustible 不燃性物质；不燃性的
incoloy 耐热铬镍铁合金
incoming line 进线口，输入线
incompact 松散的，不紧密的，不细致的，不结实的
incompatibility 不相容，不协调，不一致
incompatible 不相容的，禁忌的，不协调的
incompetent 无能力的，不合适的，不适当的
incompressibility 不可压缩性
incompressible 不能压缩的
incondensable 不能凝缩的，无凝缩性的
inconductivity 无传导性，无电导率，无电导性，不电导性

inconel 铬镍铁合金
inconformity 不适合，不一致
incongruence 不一致，不协调
incongruent 不一致的，不协调的，不和谐的
incongruity 不协调，不一致，不适宜
incongruous 不协调的，不一致的，不和谐的
inconsequent 不合理的，矛盾的
inconstancy 易变，不定性，不规则，非恒性
inconstant 多变的，无规则的
inconvertibility 不可逆性
inconvertible 不可逆的
incorrodible 不腐蚀的，不锈的，抗腐蚀的
incorrosive 不腐蚀的
incorruptibility 坚固性，耐用性，不腐败性
incorruptible 不易腐蚀的
incrassate 增厚的；浓缩，使浓厚
increase 增加[大，长]
incrustation 水锈
incurvate 向内弯曲的；使……向内弯曲
incurvation 内曲，弯曲
incurve 内弯，使……向内弯曲
indecomposable 不可分解的，不可分的
indent 缩进，订货单，凹痕；切割成锯齿状
indentation 压痕，刻痕，凹陷，缩排，呈锯齿状
indentation hardness 压痕硬度
indenter 硬度计压头
indenting 缺口，刻痕，压痕；成穴的
index 指数，索引，指针
index card 索引卡片
index signal 指示信号
indexer 指数测定仪，分度器
indicate 指示，表明
indicated defect area 缺陷指示面
indicated defect length 缺陷指示长度

indicator constant 指示剂常数
indicator electrode 指示电极
indicator paper 试纸
indicial 指数的，单位阶跃的，指标的
indicatrix 指示量，指示线，特征曲线
indifferent electrode 无关电极，中性电极
indiffusible 不扩散的
indigo 靛蓝，靛蓝染料，靛蓝色；靛蓝色的
indigotic 靛蓝的，靛青的
indirect export 间接出口
indirect hot air oven 间接加热式热风烘干室
indirect import 间接进口
indirect magnetization 间接磁化
indirect magnetization method 间接磁化法
indirect scan 间接扫查
indissolubility 不溶解性，不分解性，不均匀性，永久性
indissoluble 不能分解的，不能溶解的
indissolvable 不溶解的，难溶的
indium 铟
individual 个体的，个别的
indivisible 不可分割之事物，极微小物；不能分割的
indraft 吸[引，流]入，向内的气流，吸风[气]
induce 引起，感应，诱导
induced current method 感应电流法
induced electricity 感生电，感应电
induced polarization 激发极化，感应极化
inducement 诱因，诱导
inductance 电感，感应系数，自感应
inductance bridge flowmeter 感应电桥流量计
inductance coil 电感线圈
inductile 不能延伸的，低塑性，无延性的
induction 感应
induction apparatus 感应器
induction coil 感应线圈
induction furnace 感应电炉

induction hardening　感应淬火，感应硬化
induction heating　感应加热
induction heating evaporation system　感应加热蒸镀系统
inductive　诱导的，感应的
inductive coupled plasma emission spectrometer　电感耦合等谱仪
inductive effect　诱导效应
inductive time constant　电感性的时间常数
inductive transducer　感应传感器
inductively coupled plasma etching system　感应耦合型等离子体蚀刻系统
inductivity　诱导性，感应率，介电常数
industrial precipitation　工业沉淀物
industrial waste water　工业废水，工业污水
indurate　硬化
induration　硬[化]，固结
indurative　硬化性的，变硬的，硬结的
inelastic　非弹性的，缺乏弹性的
inelastic scattering　非弹性散射
inelasticity　不适应性，坚硬性，无弹力
inequigranular　不等粒状的
inert　惰性的，无效的
inert anode　不溶性阳极，惰性阳极
inert electrode　惰性电极
inert resin　惰性树脂
inert working electrode　惰性工作电极
inexpansibility　不可膨胀性
inextensibility　不能伸展，无伸展性
inextensional　非伸缩的
inextractable　不可萃取的
infective dose　无效剂量
inference　推论，推断
infiltrant　浸渍剂
infiltrate　渗透物；渗入
infiltration　渗入，浸润
infiltration water　过滤水
infinite　无限的，无穷的
infinitesimal　无限小的，无穷小的
infinity　无穷大，无限

inflame　燃，着火
inflammability　易燃性，可燃性
inflammable　可燃的，易燃的
inflatable　膨胀的，可充气的
inflate　使充气，膨胀，充气
inflation　膨胀，充气，打气
inflator　充气机
inflect　弯曲，使向内弯曲
inflective　弯曲的
inflexibility　缺乏弹性，刚性，不可压缩性
inflexible　不可弯曲的，不可伸缩的，刚性的
inflexion　弯曲，挠曲，向内弯曲
inflow　流入，吸入，进气
influence　影响，感应
influent　进水，进入液
influx　流入，注入，灌注
influxion　流入，注入
information　情报，资料，消息，数据
information generator　信息发送器
information storage unit　信息存储器
infra-　【构词成分】下，低于，内，间
infrared　红外线的，红外线
infrared absorption spectroscopy　红外线吸收光谱学
infrared annealer　外线退火处理机
infrared drying　红外线干燥
infrared detector　红外线探测器
infrared drier　红外线干燥器
infrared equipment　红外线设备
infrared heater　红外线加热器
infrared imaging system　红外成像系统
infrared laser　红外激光器
infrared oven　红外线干燥室
infrared photography　红外摄影术
infrared radiator　红外线辐射器
infrared ray　红外线
infrared reflection adsorption spectroscopy　红外反射吸附光谱法
infrared reflection adsorption spectroscopy

（IRRAS） 红外反射吸收光谱法
infrared sensing device 红外传感装置
infrared spectro-electrochemistry (IR-SEC) 红外光谱电化学
infrared spectrophotometer 红外分光光度计
infrared spectrophotometry 红外分光光度法
infrared spectroscopy 红外吸收光谱法
infrared spectrum 红外光谱
infrared thermography 红外热成像法，红外线温度记录法
infrasonic frequency 次声频
infrequent 稀有的，不常见的
infundibular 漏斗形的
infusibility 不熔性（熔点高于1500℃），难溶性
infusible 不熔化的，不熔化性的，难熔的
infusion 浸泡［入］，浸渍
ingot 锭，铸块
ingot mold 钢锭模，铸模
ingotism 巨晶，钢锭偏析
inhalant 吸入剂，吸入器；吸入的
inhalation 吸入（剂，物，法）
inherent 生来的，固有的，先天的
inherent fluorescence 固有荧光
inherently 天生的，固有的
inheritance 遗传，继承
inhibition 抑制，延迟，阻滞，禁［阻］止
inhibitive 抑制的，有阻化性的
inhibitor 阻化剂，防［缓］蚀剂，抑制剂
inhomogeneity 不均一，多相（性），不同类，不同质，不同族
inhomogeneous 不纯的，不均匀的，非同质的，不同类的
in-house 自身的，内部的
initial 初始，开始的，最初的
initial capacity 初始容量
initial charge 初始充电
initial data 原始数据
initial discharge capacity 起始容量
initial permeability 起始磁导率
initial state 初态
initiator 创造人，引发剂
inject forming 注射成型，注塑
injection 注射，喷射
injection molding 射出成形，喷射造型法
injection mould 注塑模具
injection pressure 注入压力，注射压力，喷注压力
injection speed 注入速度
injector pump 注射泵
injury 伤，损伤，损害
inlay 嵌体，嵌入
inlead 引入线
inlet 入口，入线，输入
in-line-of-draw 直接脱模
inner 内部的
inner lead 内部引线
inner helmholtz plane 内亥姆霍兹面
inner hexagon screw 内六角螺钉
inner layer 内层
inner potential 内电势
inner salt 内盐
innocuous 无害的，良性的
innoxious 无毒的，无害的
innovation 革新，改革
innumerable 无数的，数不清的
inoculation 孕育（作用，处理），变质处理
inordinate 无次序的，无规则的，紊乱的
inordinate wear 异常磨损
inorganic 无机的
inorganic pigment 无机颜料
inoxidizability 抗氧化性，耐腐蚀性
inoxidizable 不能氧化的，耐腐蚀的
inoxidize 使……不受氧化
inoperative 无效的，不工作的
inorganic 无机的
inorganic chemistry 无机化学
inorganic materials 无机材料

inorganic pigment 无机颜料
inorganic zinc-rich paint 无机富锌漆
inosculation 吻合，联合
in-out box 输入-输出盒
in-phase 同相，同步
in series 串联
input 输入，输入电路
input buffer 输入缓冲器
input coupler 输入耦合器
input device 输入装置
input filter 输入滤波器
input impedance 输入阻抗
input output adapter 输入-输出衔接器
input tranformer 输入变压器
insecure 不安全的，不稳定的
insequent 斜向的
insert 插入物，垫圈，插入，植入，嵌件
inserted coil 插入式线圈
inserted die 嵌入式凹模
insertion 插入
inscription 标题，注册
inside 内部，内侧
inside coil 内部线圈
inside-out 里面向外翻的
inside-out testing 外泄检测，泄出检测
inside heating method 内部加热法
insignificant 无意义的，轻微的
insolation 曝晒，日照
insolubility 不可溶性，不溶性，不溶解性
insolubilize 不溶解，降低可溶性［溶解度］
insoluble 不溶的
insoluble anode 不溶性阳极
inspection 验收，检查，商检
inspection certificate 检验证明书
inspection machine 检验设备
inspection medium 检查介质，检验介质
inspection specification 检验规范
inspection standard 检验标准
inspector 测定器，检验员
inspiratory 吸入的，吸气的

inspissant 浓缩剂；使蒸浓的，使浓缩的
inspissate 使浓缩，浓缩
inspissation 浓缩，蒸浓［浓缩］法
inspissator 浓缩器，蒸浓器
instability 不稳定性
instable 不稳定的，不牢固的
install 安装，装置
installation 安装，装置，设备
installation fundamental circle 安装基准圆
instance 例证，实例，情况
instantaneous 瞬时的
instantaneous current efficiency 即刻［瞬时］电流效率
instantaneous value 瞬时值，即时值
instead 代替，更换
instil 逐渐灌输，使渗透，滴注
instillation 滴注法，灌注
institute 学会，协会，研究所
instruction 指示，命令，说明，
instruction counter 指令计数器
institution 机关，机构，学校，制度说明书
instrument 仪器，器械，仪表
instrument air 仪表气源
instrument board 仪表板
instrument cabinet 器械柜
instrument carriage 器械车
instrument case 器械箱
instrument cover 仪器外表
instrument for nondestructive testing 无损检测仪
instrument light 仪表信号灯
instrument lubricant 器械润滑剂
instrument repairing table 器械修理台
instrument rack 器械架，仪器架
instrument stand 仪器架
instrument table 器械台，器械桌
instrumental analysis 仪器分析法
instrumental error 仪器误差
instrumentation 器械，设备，器械操作法
insuccation 浸渍

insulance 介质电阻，绝缘电阻
insulant 绝缘材料
insular 孤立的，像岛似的，隔绝的
insulate 绝缘，使绝缘，使隔热
insulated cable 绝缘电缆
insulated layer 绝缘层
insulated sleeve 绝缘套管
insulating 绝缘的，隔热的
insulating oil 绝缘油
insulating tube 绝缘管
insulation paste 绝缘胶
insulation resistance 绝缘电阻
insulativity 绝缘性，比绝缘电阻
insulator 绝缘体
insullac 绝缘漆
insusceptible 不受……影响的，不接受……的
inswept 前端狭窄的，尖端狭窄的，流线型的
intact 完整的，未受损伤的
integral 积分（的），完整的
integral capacity 积分电容
integral colored anodic oxide coating 整体着色阳极氧化膜
integraph 积分仪
integrated circuit 集成电路
integrated circuit storage 集成电路存储器
integrating instrument 积分仪，积算仪表
integrator 积分仪
integrogram 积分图
integronics 综合电子设备
intensify 使加强，使强化，增强，强化
intensity output 声强输出
intensive 加强的，集中的，重点的
intercoating adhesion 涂层间附着力
intercoating adhesion failure 涂层间剥落
inter- 【构词成分】在中间，内
interaction 相互影响，相互作用
interal stress 内应力
interaxial 轴间的
intercalation reaction 嵌入反应，插层反应
intercell connector 连接线，连接条
interception 相交，折射
interchanger 交换器
intercondenser 中间电容器，中间冷凝器
intercooler 中间冷却器
intercoupling 相互耦合，寄生耦合，相互作用
intercrescence 并生，共生，（晶体）附生
intercritical hardening 亚温淬火
intercrystalline corrosion 晶间腐蚀
interdiction 禁止，制止
interdiffusion 相互扩散
interdigital electrode array 对插梳型组合式电极
interelectrode distance 极间距
interface 分界面，界面，接口
interface boundary 界面
interface echo 界面回波
interface trigger 界面触发
interfacial （晶体或其他固体）面间的，面际的，界面的，形成界面的
interfacial tension 界面张力
interfacial potential difference 界面电势差
interfacial tensiometer 界面张力计
interference 干涉
interference absorber 干扰吸收器，干涉滤光片
interference filter 干涉滤波器，干涉滤光镜
interference microscope 显微镜干涉法，干涉显微镜
interference preventer 防干扰装置
interference refractometer 干涉折射计
interference signal 干扰信号
interference spectroscope 干涉分光镜，干扰光谱仪
interferogram 干涉图
interferometer 干涉仪，干扰计
interferoscope 干涉镜，干扰显示器
intergranular corrosion 晶间腐蚀，粒间

腐蚀
intergranular fracture 晶间断裂，沿晶断裂
intergranular stress corrosion cracking 晶间应力腐蚀断裂
interlayer 夹层，隔层
interspace 空间，间隙，中间
intermediate 中间物，介于
intermediate frequency 中频
intermediate solid solution 中间固溶体，次生固溶体
intermetallic 金属间化合物；金属间（化合）的
intermetallic compound 金属间化合物
intermiscibility 相溶性，互溶性
intermission 间断，间歇
intermittent discharge 间歇放电
intermittent feed 间歇投药，间歇进给
intermodulation 互调，相互调制
intermodulation voltammetry 互调制伏安法
intermolecular forces 分子间作用力
internal abstraction 内夺取
internal carbon referencing 内标碳参考
internal conversion 内转换
internal electrolysis 内电解法
internal energy 内能
internal plasticization 内部塑化
internal porosity 内部气孔
internal reflection spectroelectrochemistry (IRS) 内反射光谱电化学
internal scattering 内散射
internal standard 内标法
internal stress 内应力
international standard 国际标准
international standard organization 国际标准化组织
interphase 界面
interpolation 插入，内插法
interpolymer 交聚物
interpretation 翻译，解释，说明

interrupt 断续，中断
intersection 相交，交叉
interstage amplifier 级间放大器
interstice 间隙原子，裂缝，空隙
interstitial 填隙原子，节间；间质的，空隙的，填隙的
interstitial diffusion 间隙式扩散
interstitial oxygen 晶格间氧气，格隙氧气
interstitial solid solution 间隙固溶体
interstitial velocity 空隙速度
intersystem crossing 体系间跨越
interval 间隔，时间间隔，中断期
interval between coats interval 涂装间隔
interval timer 限时器
intervalometer 定时器，时间间隔计
intra- 【构词成分】在内部
intrasonic 超低频
intrinsic 本质的，固有的
intrinsic semiconductor 本征半导体
introduce 引进，引导，提出
introduction 说明书，前言，绪论
introscope 内腔检视仪，内孔窥视仪
intumescent additive 膨胀剂
invagination 凹入，折入，套叠
invariant 不变的
invariant point 不变点
invariant zone 不变区
invention 发明，创造
inventory 清单，存货单
inversion 转化，倒置，反向，倒转
inverted region 翻转区，反转点
invest 包埋，围模，附于
investigation 调查，研究
investment casting 精密铸造，熔模铸造
invisible spectrum 不可见光谱
involuntary 不随意的，偶然的
involve 包含，包括
iodide 碘化物
iodimetry 碘量法
iodine 碘
iodine value 碘值

ion abundance 离子丰度
ion analyser 离子分析仪
ion beam analysis (IBA) 离子束分析
ion beam etching system 离子束蚀刻系统
ion beam induced mass transport 离子束诱导质量迁移
ion beam probe 离子束探针
ion beam sputtering system 离子束溅镀系统
ion bombardment 离子轰击
ion carbonitriding 离子渗碳氮化
ion carburizing 离子渗碳处理
ion chromatography 离子色谱法
ion concentration cell 离子浓差电池
ion crystal structure 离子晶体结构
ion dipole force 离子偶极力
ion exchange 离子交换
ion exchange chromatography (IEC) 离子交换色谱法
ion exchange membrane 离子交换膜
ion exchange resin 离子交换树脂
ion implantation 离子植〔注〕入，离子移植
ion injecting 离子注入
ion micro probe analysis method 离子微探针分析法
ion nitriding 离子渗氮
ion pair 离子对
ion plating 离子镀
ion selective electrode (ISE) 离子选择性电极
ion source 离子源
ion suppression chromatography (ISC) 离子抑制色谱法
ion-scattering spectrum 离子散射谱
ionic activity 离子活度
ionic bond 离子键
ionic conductor 离子导体
ionic crystal 离子晶体
ionic equation 离子方程式
ionic exchange 离子交换

ionic polarization 离子极化
ionic radius 离子半径
ionic reaction 离子反应
ionic solid 离子晶体，离子固体
ionic strength 离子强度
ionization 电离，游离，离子化
ionization constant 电离常数
ionization energy 电离能
ionization vacuum gage 电离真空计
ionization potential 电离电位
ion-nitriding 离子氮化
ionogram 电离图
ionography 离子放射照相法
ionosphere 电离层
ionotron 静电消除器
ipso position 本位
IR oven 红外线烘烤炉
iron phosphating coating 铁盐磷化膜
iridescence 干涉色，彩虹色
iridium 铱
iron phosphating coating 铁盐磷化膜
irradiation 照光，辐射
irregular 不规则的，无规律的
irregular lighting 不规则照明
irreversibility 不可逆性
irreversible 不可逆的
irreversible process 不可逆过程
irreversible wave 不可逆波
IR spectrophotometer 红外线分光光度计
iso- 【构词成分】同，等，均匀
isobutyl 异丁基
isochrone 等时线，瞬压曲线
isoelectric point 等电点
isoenergetic electron transfer 等能电子转移
isoforming 异构重组，异构重整
isolated 孤立的
isolation 隔离度，绝缘
isolation room 隔离室
isolator 绝缘体，隔离器，隔离物
isomer 同分异构体，异构物

isomerism 同分异构现象，异构现象
isomorphous 同形的，同晶的
isopotential 等电势[位]的，等电的
isopropyl 异丙基
isopropyl alcohol 异丙醇
isotactic 全同立构的
isostatic pressing 均压法，等压压制
isotherm 等温线
isothermal annealing 等温退火
isothermal line 等温线
isochronism 等时性
isothermal 等温的
isothermal quenching 等温淬火
isothermal transformation diagram 等温转变相图
isotope 同位素
isotope tracer 同位素示踪物
isotope X-ray fluorescence spectrometer 同位素X荧光光谱仪
isotropic 各向同性的
isotopic ion 同位素离子
isotopic tracer 同位素指示剂，示踪原子
isotropic etching 各向同性蚀刻，等向性蚀刻

J

jack 千斤顶
jackbit 钻头
jacket 套，给……装护套
jacket water 水套冷却水
jackmanizing 深渗碳处理，深度渗碳
jade 翡翠，碧玉
jag 缺口，狂欢；使成锯齿状，使成缺口
jagged 锯齿状的，参差不齐的
jam 使堵塞[塞满]，混杂，压碎；堵塞，轧住
jammed 堵塞的，拥挤的，轧住了的
jamming 干扰，堵塞，抑制
jamming rate 干扰率
jamp up 孔眼堵塞
jamproof 抗干扰的
jar 罐，广口瓶，震动，刺耳声；冲突，不一致，震惊，发刺耳声，震动，刺激，使震动
jarring 碾轧声，冲突，震动；不和谐的，刺耳的，辗轧的
jarring machine 冲击机，震动机
jaw 钳子，量爪，凸轮，夹板，游标
jejune 枯燥无味的，空洞的，内容贫乏的
jell 使……成胶状，使……定形
jellied 涂凝胶物的，凝成胶状的
jellification 胶凝，冻结
jellify 使成凝胶状
jelly 胶状物；成胶状，使结冻
jet 喷口，喷嘴；射出
jet dryer 喷射干燥器
jet electro-plating 喷射电镀
jet hardening 喷射淬火，喷射淬硬法
jet quenching 喷射淬火，喷射淬硬法
jet test 喷射试验，喷流实验
jetter 喷洗器
jettison 投弃，投弃货物，下坠
jetty 伸出，突出
jig 夹具

jigger rotor 盘车转子
jigging 筛，振动，上下簸动
jiggly 不稳定的，摇晃的
jitter 振动，晃动，(信号的)不稳定性
jitterbug 图像不稳定
job site 施工现场
jog 割阶，粗糙面
jogged 嵌合的，拼合的
jogged dislocation 割阶位错
jogged screw dislocation 阶梯状螺旋型位错
joggle 啮合，轻摇，摇动，榫接；啮合，摇动，摇曳，轻摇
joint 接合，连接
joint line 接缝，合模线，结合线
jointing 填料，焊接
joist 工字钢
joist shears 型钢剪切机
joist steel 梁钢
jolt 颠簸，摇晃，震惊，严重挫折；使颠簸；
jominy curve 顶端淬火曲线
jominy distance 顶端淬火距离
jominy test 顶端淬火试验
jounce 震动，颠簸；使颠簸，使震动
journal 日报，杂志，日记，分类账
judder 颤抖，声调急剧变化；颤抖，颤动
jug 水壶，水罐，汽缸
juicy 多汁的，利润多的
jumble 混乱，杂乱的一堆东西；混[搀]杂，使混乱
jumbly 混杂的，混乱的
jump frequency 跳变频率，跃迁频率
jumper 跳线，工作服
junction 接头，连接点
junction box 接线盒，分线箱
junction transistor 面结型晶体管
juncture 接缝，连接，接合

junk 废料，垃圾
jury 应急的
just etching 适量蚀刻
justice 法律制裁，正义，公正，合理

jut 突出部分，尖端；突出，伸出
juxtapose 并列，并置
juxtaposition 并置，并列，毗邻

K

kaleidoscopical 千变万化的
kali 氧化钾
kali salt 钾盐
kalimeter 碳酸定量器，碱定量器
kalium 钾
kalk 石灰
kalsomine 白粉溶液，刷墙粉；粉刷
kampometer 热辐射计
kaoline 高岭土
karbate （耐蚀衬里材料）无孔碳
katabatic 下降的，下吹的
katabolism 分解代谢
Katalysis 催化作用
kataphoresis 电泳，电粒降泳
kathode 阴极，负极
kation 阳离子，正离子
katogene 破坏作用
katogenic 分解的
kaurit 尿素树脂接合剂
kelvin 绝对温度，开尔文温度
kerf 切口，截口，劈痕
kerf loss 截口损失
kerf thickness 刀刃（截口）厚度
kern 颗粒，（铅字面之）上下的突出部分；使铅字上下突出，将……做平
kernel 核心，要点，精髓
kerosene 煤油
ket- 【构词成分】酮
ketal 缩酮，酮缩醇
keto- 【构词成分】酮
ketogenesis 生酮作用，酮生成
ketolytic 解酮作用的
ketone （甲）酮
ketone hydrate 酮水合物
ketones solvent 酮系溶剂
ketonic 酮的
ketonization 酮基化作用
ketonize 酮化

kevlar fibre 芳纶纤维
key ingredients 关键成分
keyway 键槽，汽缸闸之栓孔，凸凹缝
kick 反冲
kickback 反冲，逆转，退还
kicksort 脉冲幅度分析，振幅分析
kieselguhr 硅藻土
killed steel 脱氧钢
kiln 砖，石灰等的）窑，炉，干燥炉；烧窑，在干燥炉干燥
kilo- 【构词成分】千
kilovolt 千伏特
kilogram 公斤
kindling 点火，引火物
kinematical 运动学的
kinematical viscosity 运动黏度
kinetic 运动的，活跃的，动力（学）的
kinetic acidity 动力学酸度
kinetic control 动力学控制
kinetic current 动力电流，动力波
kinetic energy 动能
kinetic equation 动力学方程
kinetic viscosity 动力黏（滞）度
kinetic wave 动力波
kinetics 动力学
kinetics control 动力学控制
kinetics of dissolution 溶解动力学
kink 扭结；使扭结
kink band 扭折带
knead 揉合，揉捏，捏制，混合，搅拌
knife 劈开
knife-line corrosion 刀线腐蚀
knifing 刮涂
knit 编织，结合，合并，弄紧，弄结实，使紧凑
knockout 脱模装置
knob 把手，瘤，球形突出物；使有球形突出物

knobbed 有节的，有圆头的
knobble （树的）小节，小瘤，小疙瘩
knot 结，节瘤，疙瘩；打结
knotty 棘手的，难解决的；多节的，瘤状的

knurl 隆起，节，瘤，硬节
knurling 滚花，压花纹，滚花刀，压花刀
krypton 氪
kurtosis 峭度，峰态

L

labelled atom　示踪原子
labelled　贴上标签的，标记的，示踪的
label　标签，商标，签条；标注，贴标签于
labile　易变的，不稳定的，不安定的
lability　易变性，不稳定性
labilized hydrogen atom　活化的氢原子
labilize　使不稳定，活化
labyrinth　错综复杂，复杂的事
lac varnish　虫漆清漆
lacerable　撕得碎的，划得破的
laceration　裂伤，撕裂，割破
lace　精细网织品，花边；束紧
lacing　束紧，导线
laciniate(d)　有边的，穗状的，成锯齿状的
lack of painting　烤漆不到位
lacklustre　无光泽，无生气；无光泽的，无生气的
lack　缺乏，不足
lacquer enamel　挥发性磁漆，珐琅漆
lacquer solvent　溶漆剂，助溶剂
lacquer thinner　挥发性漆稀释剂
lacquer　漆；涂漆，使表面光泽
lacquering　上漆，涂漆层
lacquerless　无漆的
lactam　内酰胺
lactate　乳酸盐
lacteal　乳状的，含乳状液的
lacteous　乳状的，乳白色的
lactescence　乳状液，乳汁色
lactic acid　2-羟基丙酸，乳酸
Lactic　乳的，乳汁的
lactol　内半缩醛，乳醇
lactone　内酯
lacunal　有孔的，空隙的
lacunary　有孔的，缺项的，有缺陷的
lacunose　多孔的，多空隙的
lac　虫胶，虫漆

lade　装载，汲出，获得，塞满
laggard　落后的，迟钝的，迟缓的
lagging casing　保温外壳；绝热外壳；绝热外壳
lagging coil　滞后线圈
lagging commutation　滞后换向；延迟换向
lagging curve　滞后曲线
lagging device　滞后装置，滞相装置
lagging edge　延迟反馈，后沿，下降边
lagging effect　滞后效应
lagging feedback　迟滞反馈
lagging filter　迟后滤波器
lagging half axle　横半轴
lagging index　滞延指数
lagging jacket　汽缸保温套
lagging load　电感性负载，滞后负荷
lagging material　绝热材料，绝热材料
lagging phase angle　滞后相位角
lagging phase　滞后相位
lagging power factor　滞后功率因数
lagging voltage　滞后电压
lagging　绝缘层材料；落后的
lag　迟延；最后的；滞后，缓缓而行
lamellar air-heater　片式热风器
lamellar body　片层体，层状小体，板层小体
lamellar compound　层状化合物
lamellar crystal　片状晶体
lamellar crystallization　片状结晶
lamellar deformation　片晶变形
lamellar displacement　片晶位移
lamellar eutectic　层状共晶
lamellar extrusion technique　层状挤压成形技术
lamellar flow　片流，层流
lamellar fracture　层状断裂面，层状断口
lamellar graphite　片状石墨
lamellar growth　片状生长

lamellar lattice 片晶晶格
lamellar martensite 片状马氏体，层状马氏体
lamellar membrane 片层膜，片状膜
lamellar morphology 片状形态
lamellar pearlite 片状珠光体，层状珠光体
lamellar phase 层状相
lamellar spacing 层间隙
lamellar structure 层状组织，薄层状结构，片层结构，片晶结构，片状组织，带状组织
lamellar tearing 层状撕裂，层间撕裂
lamellar thickening 片晶增厚
lamellar toughening 层状化增韧
lamellar 薄片状的，薄层状的
laminar boundary layer 层流边界层
laminar composite 层状复合材料
laminar composite 层状金属复合材料
laminar convection 分层对流
laminar crack 层状裂纹
laminar flow 层流
laminar mixing 分层拌和
laminar region 层流区
laminar structure 层性结构，薄片状构造
laminar symmetry 分层对称性
laminar tissue 薄板组织
laminar 层状的，薄片状的，板状的
laminating method 被覆淋膜成形
lancing 清除
lanthanide 镧系（元素）
lap width 搭接宽度
lapped 重叠的，互搭的
lapper 研磨机
lapping 抛光/研磨
lapse line 递减线
lapse rate 减率，直减率，递减率
lapse 失效；失效，失检
large acreage 大面积
large air compressor 大型空气压缩机
large amounts 大批

large amplitude non-linear condition 大振幅非线性条件
large amplitude shock wave 大波幅激波
large amplitude 大振幅
large angle boundary 大角度晶间界
large angle 大角度
large aperture interferometer 大孔径干涉仪
large aperture 大孔径
large area colouring 大面积着色
large batch production 大批生产
large batch 大批量的
large blocked structure 大块状结构
large capacity meter 大流量计
large capacity refrigerated centrifuge 大容量冷冻离心机
large deformation 大变形
large deviation 大偏差
large diameter 大直径
large elastic deformation 大弹性形变
large grained 大颗粒，大晶粒
large output 大量生产，大输出
large particle composite 大颗粒复合材料
large power 大功率
large quantities 大批
large quantity 大量
large scale 大比例尺，大尺度，大规模的
large-angle scanning 广角扫描
large-scale 大批的，大型，大比例尺的，大规模，大规模的
large-type horizontal metallurgical microscope 大型卧式金相显微镜
laser cutting 激光切割
laser electroplating 激光电镀
laser engraving 激光雕刻
laser glazing 激光釉化
laser heat treatment 激光热处理
laser surface improving 激光表面改性
laser 激光
latching 封闭，封锁
latch 闩锁，闭锁

latence	潜在，潜态
latent image	潜象
latent	潜在的，潜伏的，隐藏的
lateral	侧部；侧面的，横向的
lateral scan with oblique angle	斜平行扫查，宽容度
lateral scan	左右扫查，横向扫描
lateral translation	侧向平移
laterally	旁边地，在侧面，横向地
latericeous	有砖之红色的，如砖的，土红色的
laterodeviation	侧向偏斜，侧偏
latero-	【构词成分】侧，旁
latexometer	胶乳比重计
latex	乳胶，乳液
lathe	车床
lathe cutter	车床切削刀具，割板机
lathe cutting tool	车床切削刀具
lathe grinding	车床磨削，车床研磨
lathe lapping	车床研磨
lathe turning	车床车削；车床切削
lather	肥皂泡，激动；涂以肥皂泡，起泡沫
lathy	细长的，板条似的
latticed	有格子的，制成格状的，格子形的
laticiferous	含乳液的，有乳液的
laticometer	乳胶比重计
lati-	【构词成分】宽的，阔的
lattic defect	晶格缺陷
lattice constant	晶体常数；点阵常数
lattice parameter	晶格参数
lattice strain	晶格应变
lattice	晶格，点阵
latticing	成（网）格状，缀合
latus	侧，边，弦
launder	流水槽；洗涤
lauricacid	十二酸，月桂酸
lauroyl peroxide	月桂酰过氧化物
lauroyl	月桂酰，十二烷酰酰基
lauryl alcohol	十二醇，月桂醇
lauryl aldehyde	月桂醛
lauryl amine	月桂胺，十二胺
lauryl mercaptan	月桂硫醇，月桂硫醇
lauryl sodium sulfate	十二烷基硫酸钠
lauryl sulfate	十二烷基硫酸，月桂烷硫酸酯
lauryl	十二（烷）基；月桂基
lavation	洗涤，洗涤用的水，洗去法，冲洗法
lavender	薰衣草，淡紫色；淡紫色的
lavish	浪费的；滥用
layer corrosion	层状腐蚀，层间腐蚀
layer	层，叠片；把……堆积成层
laying-off	下料，停工
layout drawing	布置图，配线图，草图
leach	过滤，过滤器；过滤，萃取，被过滤
leach away	滤除，过滤掉
leach out	滤去，浸出，渗漏，淋溶
leachable	可滤取的
leaching solution	浸提液
leaching	沥滤，淋洗，浸出，洗盐
lead	引线，连接线；铅；衬铅
lead accumulator	铅蓄电池
lead acetate paper	乙酸铅试纸
lead acetate poisoning	醋酸铅中毒
lead acetate	醋酸铅，乙酸铅
lead acid batteries	铅酸电池
lead acid storage battery	铅酸蓄电池
lead annealing	铅浴退火
lead antimonate	锑酸铅
lead apron	铅防护板
lead bath quenching	铅浴淬火
lead bath treatment	铅浴处理
lead calcium batteries	铅钙电池
lead chloride	氯化铅
lead chromate	铬酸铅
lead corrosion test	铅腐蚀试验
lead deposit	铅沉积
lead dichloride	二氯化铅
lead dichromate	重铬酸铅

lead dioxide 二氧化铅
lead fluorosilicate 氟硅酸铅
lead fuse 铅保险丝
lead halide 卤化铅
lead hydroxide 氢氧化铅
lead lining 铅衬，铅衬里，铅内衬
lead orthoplumbate 四氧化三铅
lead paint 铅丹，铅漆，铅涂料
lead patenting 铅浴等温淬火，铅浴索氏化处理
lead peroxide 二氧化铅，过氧化铅
lead pigment 铅系颜料
lead plating 镀铅
lead protoxide 一氧化铅
lead quenching 铅浴淬火
lead screen 铅屏，铅增感屏
lead shielding 铅屏蔽
lead silicofluoride 硅氟化铅，六氟硅酸铅
lead storage battery 铅蓄电池
lead sulf(ph)ate 硫酸铅
lead sulfide 硫化铅
lead tank 铅衬槽
lead titanate 钛酸铅
lead-acetate test 乙酸铅试验
lead-antimony alloy 铅锑合金
leading peak 前沿峰，谱带伸长
leading wire terminal 引线端子
leaf separator 叶片式隔板
leafing 漂浮
leak 泄漏
leak detector 检漏仪
leak test 泄漏试验
leakage field 泄漏磁场，泄漏场
leakage 泄漏
leakproofness 密封性，严密性
leak-tight 防漏的，气封的，水密的，密封的
leaky 漏的，有漏洞的
leaving group 离去基团
lecithin 卵磷脂
ledeburite 莱氏体

ledge 突出的部分
leftover 残余的；剩余物，废料
legend 说明，图例，图表符合，代号，说明书
lemon 柠檬色的，淡黄色的
lengthen 使延长，加长；延长，变长
lengthways 纵向的；纵向地
length 长度，距离，截距
lensed 有透镜的
lens 镜头，透镜
lenticular 透镜的，两面凸的
lentic 死水的，静水的
lentoid 透镜状结构；透镜状的
leptokurtosis 峰态，峭度
leptopel 微粒，胶质
lepto- 【构词成分】小，细，薄
lethal 致命的
leuco 白色，无色
leuc- 【构词成分】无色，白，淡
level gauge 液位计
level instrument 位面计，水平仪
level switch （信号）电平开关
level 水平
leveling action 整平作用
leveling agent 均化剂，匀染剂
leveling 流平性，整平，校平
lever rule 杠杆定理
levo- 【构词成分】左旋
liberalize 使自由化，解除对……的控制
librate 摆动，平均
libratory 摆动的，振动的
license 许可证，执照
lifting 举起，起重，提高，上升
ligand 配位体
ligate 绑，扎
light activated 光敏的
light aging 光致老化
light alloy 轻合金
light color 浅色
light curing coating 光固化涂料
light intensity 光强度

light scattering method 光散射法
light stabilizer 光稳定剂
light 淡（浅）色的，轻的
lightfast 耐晒的，不褪色的
lightless 不发光的，无光的，暗的
lightproof 不透光的，遮光的
ligroin 轻石油，挥发油，石油醚
lilaceous 淡紫色的
liliquoid 乳状胶体
limes 边界，界限
liminal value 极限值
liminal 阈限的，极限的
limit of explosion 爆炸极限
limitation 限度，局限性
limited charge voltage 充电限制电压
limited voltage in charge 充电上限电压
limiting concentration 极限浓度
limiting current density 极限电流密度
limiting current 极限电流
limiting resolution 极限分辨率
limiting wear 极限磨损
line defect 线缺陷
line focus 线焦点，行聚焦
line pulling 线拉伸
line scan 行扫描
line scanner 线扫描器，行扫描仪
line spectra 线光谱
line streching 线拉伸
linear attenuation coefficient 线性衰减系数
linear coefficient of thermal expansion 线性热膨胀系数
linear polymer 线型聚合物
linear potential sweep chronoamperometry 线性电势扫描计时电流法
linear scan 线扫查，线性扫描
linear sweep voltammetry (LSV) 线性扫描伏安法
linear test system 线性测试系统
linear 线性
linearity amplitude 线性振幅
linearity distance 线性距离
liner 衬垫，衬套
linewidth 线宽，谱线宽度
lining 衬里，内层，衬套
linkage 连接，结合
linoleic acid 亚油酸，髎酸
linolenic acid 亚麻酸
liny 似线的，划线的，有皱纹的
lipoid 类脂的，脂肪性的
lipolytic 分解脂肪的，脂解的
lipophilic 亲脂［油］的
lipophilic emulsifier 亲油性乳化剂，亲油性去除剂
lipophilic group 亲油基
lipophilic remover 亲油性洗净剂
lipophilicity 亲油性
lipophobicity 疏油性
liposoluble 脂溶的
liquate 熔解，熔析，熔融
liquation 熔融，熔析，偏析
liquefacient 液化剂；液化的，溶解性的
liquefaction 液化，熔解，冲淡，稀释
liquid chromatography 液相色谱法
liquid crystal 液晶
liquid honing 液体喷砂法
liquid junction boundary 液接界面
liquid junction potential 液接电位
liquid membrane permeation (LMP) 液膜分离技术
liquid metal corrosion 液态金属腐蚀
liquid penetrant examination 液体渗透检验
liquid source delivery system 液体源输送系统
liquid spill sensor 液体溢流感测器
liquidus line 液相线
liquid-junction potential 液体接界电位
liquid-liquid chromatography 液-液色谱法
liquid-liquid extraction 液-液萃取法
liquid-solid adsorption chromatography 液固吸附色谱法

lithiation 锂化
lithium ion batteries 锂离子电池
lithium polymer batteries 锂聚合物电池
lithium 锂
litmus 石蕊
load circuit 负载电路
load test 负荷试验，加载试验
load 荷载，工作量，装载量
loading factor 负载因数
lobate 叶状的，有叶的
local action 局部作用
local corrosion 局部腐蚀
local diamagnetic shielding 局部抗磁屏蔽
local distortion 局部变形，局部畸变
local electric field effect 局部电场效应
local heat treatment 局部热处理
local magnetization 局部磁化
local scan 局部扫查
localization index 定位指数
localized electrochemical deposition (LECD) 局部电化学沉积
location accuracy 定位精度，定位准确度
location marker 定位标记
location 定位
lodgement 沉积，沉淀，堆积物
loft-dried 风干的
logarithmic 对数
logic diagram 逻辑图
logistics flow 物流流程
logy 迟缓的，弹性不足的
lone pair electron 孤对电子
long range shielding effect 远程屏蔽效应
longitudinal field 纵向场
longitudinal magnetization method 纵向磁化法
longitudinal resolution 纵向分辨率
longitudinal 纵向的
longitudinal 长度的，纵向的，经线的
loop tenacity 环结强度
loop test 环路测试
loosely packed 疏堆积的

loss factor 损耗系数
loss of gloss 失光
low alloy steel 低合金钢
low density polyethylene (LDPE) 低密度聚乙烯
low pressure heat treatment 低压热处理
low pressure steam 低压蒸汽
low rate discharge characteristics 低率放电特征
low temperature annealing 低温退火
low temperature superconducting material 低温超导材料
low temperature tempering 低温回火
low-carbon steel 低碳钢
low-energy electron diffraction (LEED) 低能电子衍射
low-energy ion-scattering spectrometry 低能离子散射谱
low-energy photon radiation 低能光子辐射
low-field approximation 低场近似
low-frequency induction furnace 低频感应炉
low stress 低应力
low-shaft furnace 坑式炉
lower critical temperature 下限临界温度
lower level line 下液面线
lubricant 润滑剂；润滑的
lubricating oil 润滑油
lug 支托，接线片；用力拉或拖
luminance 亮度，发光度
luminescence 发冷光，荧光
luminosity 亮度，发光度，光度
luminous 发光的，明亮的，清楚的
lumping 很多的，大量的
lump 块，块状；成团的，总共的
lusterless 没有光泽的
luster 光泽
lustrous 有光泽的，光辉的
luteous 略带绿色的黄金色的，深橙黄色的
luxuriant 繁茂的，丰富的，多产的

lye change 碱液；用碱液洗涤
lye dissolving tank 溶碱槽
lye 碱液
lyolysis 液解，溶剂分解，溶剂解
lyophile 亲液胶体，亲液物
lyophilic colloid 亲液胶体
lyophilic group 亲液基
lyophilic polymer 亲液性聚合体
lyophilic 亲液的
lyophilisation 低压冻干法，冰冻干燥法，升华干燥
lyophilize 使冻干
lyophilizing 冻干
lyophobic 疏液的，憎液的
lyophoby 疏液
lyosorption 吸收溶剂（作用）
lyotrope 易溶物，感胶离子
lytic 溶（松）解的

M

maceration 浸渍（作用）
machinability 可切削性，可加工性，机制性
machinability annealing 改善加工性的退火
machinable 可切削的，可加工的，可用机械的
machine coarsening 机械粗化
machine set 机组
machine shaping 加工成型
machineable 可加工的
machinery 机械，机械设备，工具
machining 机械加工
mackenite metal （镍铬系，镍铬铁系）耐热合金
mackintosh 防水胶布
macle 双晶，矿物中的暗斑
macro 宏观的，大量的，常量的；宏观
macro check 宏观分析，宏观[低倍，肉眼]检查
macro etching 宏观试片腐蚀
macro qualitative analysis 常量定性分析
macro streak flaw test 断面缺陷肉眼检查，粗视条痕裂纹检查
macroanalysis 常量分析
macroanalytic 常量分析的
macrochemistry 常量化学
macrocrystalline 宏晶的，粗（粒结）晶的，大（块）结晶的
macrocyclic 大环的
macrodispersoid 粗粒分散胶体
macroeffect 宏观效应
macroetch 宏观腐（侵）蚀，粗视组织侵蚀
macrograin 粗[宏观]晶粒
macrographic examination 宏观检查，粗视组织检查
macrographic 宏观的，低倍照相的

macroheterogeneity 宏观不均匀性
macroion 大（分子）离子，高（分子）离子
macrolide 大环内酯
macromeritic 粗晶粒状的
macromolecular 大分子的，高分子的
macromolecule 大分子链，大[高]分子
macrophotograph 放大照相[照片]
macroporosity 大孔性，宏观[肉眼]孔隙
macroporous 大孔的
macroscopic test 低倍检查
macroscopic 宏观的，低倍放大的，粗视的
macroscopical 宏观的，低倍放大的，粗视的
macrosection 宏观断面（图），粗视剖面
macrostrain 宏应变
macrostress 宏应力
macrostructure 宏观[金相]组织，粗视[低倍，肉眼可见的]组织
macrovoid ratio 大孔隙比
macro-axis 长轴，斜方晶体或三斜晶体中的长轴
macro-crack 宽[裂]缝，宏观裂缝
macro-diagonal 长对角轴的
macro-reticular typeion exchange resin 大孔型离子交换树脂
macro-structure 宏观组织
macro-throwing power 宏观分散能力
macular 有斑点的，不清洁的
maculate 弄脏，玷污
maculation 斑点，污点
magaluma 铝镁合金
magenta 深红色的，（碱性）品红；红色苯胺染料，洋红染料
magnadure 铁钡永磁合金，马格那多尔磁性合金

magnaflux 磁粉检查法
magnalium 镁铝合金
magnesia 氧化镁，镁土
magnesium carbonate 碳酸镁
magnesium 镁
magnet 磁铁
magnetic anisotropy 磁各向异性
magnetic circuit 磁路
magnetic domain 磁畴
magnetic field distribution 磁场分布
magnetic field indicator 磁场指示器
magnetic field meter 磁场计
magnetic field strength 磁场强度
magnetic field 磁场
magnetic flux density 磁通密度
magnetic flux 磁通
magnetic force 磁力
magnetic gage 磁量规
magnetic heat treatment 磁场热处理
magnetic hysteresis 磁性滞后，磁滞现象
magnetic induction 磁感应强度
magnetic intensity 磁感应强度
magnetic leakage field 漏磁场
magnetic leakage flux 漏磁通
magnetic moment 磁矩
magnetic particle examination 磁粉检验
magnetic particle indication 磁痕，磁粉显示
magnetic particle inspection 磁粉探伤法
magnetic particle inspection flaw indications 磁粉检验的缺陷，显示（缺陷磁痕）
magnetic particle test 磁粉探伤
magnetic particle 磁粉
magnetic permeability 磁导率
magnetic pole 磁极
magnetic quantum number 磁量子数
magnetic recording materials 磁记录材料
magnetic saturation 磁饱和
magnetic stirrer 磁力搅拌
magnetic storage medium 磁存储介质
magnetic storage 磁存储器

magnetic susceptibility 磁化率
magnetic-sector mass spectrometer 扇形磁场质谱仪
magnetism 磁性
magnetization 磁化
magnetization magneto-microwave plasma etching system 磁场微波型等离子体蚀刻系统
magnetizing 磁化
magnetizing current 磁化电流
magnetostriction 磁致伸缩
magnetostrictive effect 磁致伸缩效应
magnetron sputtering system 磁控管溅镀系统
magnifying glass 放大镜
main group 主族
maintenance factor 维护率
major defect 主要缺陷
make-up 补充
maleic acid 马来酸，失水苹果酸
malfunctioning 出故障的
malic acid 苹果酸，羟基丁二酸
malic amide 苹果酰胺
malleable cast iron 可锻铸铁，展性铸铁
malleable iron 可锻铸铁
malleable 可锻的，可塑的，有延展性的，易适应的
malleablizing 可锻化退火
manganese 锰
manganese dioxide 二氧化锰
manganese lithium oxide 锰酸锂
Mannich base 曼尼希碱
manometer 压力计，测压计
manufacture 制造，产品，制造业；制造，加工，捏造
manufactured 制造的，已制成的
manufacturing 制造业，工业；制造的，制造业的
mar 污点，瑕疵；损毁，损伤，糟蹋，玷污
maraging 高强度热处理，马氏体时效

Marble corrosive liquid 硫酸铜盐酸（钢铁显微组织检查用）腐蚀液，马布尔侵蚀剂
marcomizing 不锈钢表面氮化处理
margaric acid 十七（烷）酸
margaric 十七烷的
margin of stability 稳定系数
margin 边（缘，界，限，距，际）
marginal 边缘的，临界的，限界的
margination texture 蚀边结构
marine corrosion 海水腐蚀，海洋腐蚀
marine 海运的，航海的；海运业
marked 有标记的，标定的，明显的，显著的
marking 刻印加工
marquench 分级淬火，等温淬火
marquetry 镶嵌细工
marquetry work 镶嵌装饰品
martemper 使分级淬火，使等温淬火
martensite 马氏体
martensitic 马氏体的
martensitic stainless steel 马氏体不锈钢
Martin 马丁［平］炉
Martin furnace 马丁炉，平炉
Martin steel 平炉钢
Martinel steel 硅锰钢（碳 0.24%，锰 0.75%，硅 0.1%）
marvel 奇迹，奇观；对……感到惊异
marworking 形变热处理，奥氏体过冷区加工法
mash 磨碎，捣烂，混合
masher 磨碎机
masking 掩蔽，遮盖
masking agent 掩蔽剂
masking power 遮盖力
masking tape 不透光胶带，遮蔽胶带
maskless 无遮蔽的，无屏蔽的
mass 物质，质量，大量，成批
mass action 质量［浓度，分量］作用
mass analyzer 质量分析器，质谱仪
mass analyzing system 质量分析系统
mass balance equation 质量平衡式
mass data 大量数据
mass balance 质量平衡
mass diagram 积分曲线
mass flow 质量流量
mass hardness 全部过硬
mass load 惯性负载
mass number 质量数
mass production 大量生产，成品生产
mass range 质量范围
mass resistivity 比电阻，质量电阻率
mass spectrograph 质谱仪
mass spectrometer 质谱仪，质谱分析器
mass spectrometric analysis 质谱分析
mass spectrum 质谱
mass tone 主色，浓色
mass transfer coefficient 传质系数
mass transfer 物质移动，传质
massic 质量的
massicot 氧化铅，铅黄，铅丹
massive 重的，大的，大而重的，非晶质的
massive structure 整体结构，厚块结构，块状组织
master 基本的，主的，总的，仿形［校正，精通］的
master check 校正，校对
master color 造型色板
master frequency 主频
master mould 原始模型，母型
master sample 校准样品
master-slave 主从的，仿效的
mastery 精通，熟练，控制，掌握
mastic 胶，膏，树脂，胶黏剂，胶粘水泥
mat 褪光；暗淡的，无光泽的
mat coat 罩面，面层，保护层
mat finish 消光处理
mat fracture 无光泽断口
mat glass 磨砂玻璃，毛玻璃
mat metal 未抛光的金属
matchable 对等的，相配的
matchboard 型板，模板

matched die method 对模成形法
matching 匹配，双合，配合
matching impedance 匹配（用）阻抗
material 材质，材料
material balance 物料平衡
material certificate 材料合格证
material process technics 材料加工工艺
material production technics 材料生产工艺
material technics 材料工艺
material thickness 料片厚度
mathematical 数学的
mating 配合的，配套的，相连的
mating surface 啮合表面
matrix 基体，母体，基质
matrix effect 基体效应
matrix phase 基体相
matt 无光的，无光泽的，暗淡的，乌泽的，不光滑的，粗糙的；使无光泽
matt paint 无光泽涂料
matt surface 无泽面
matte （冶炼中产生的）不纯金属，褪光；无光泽的，暗淡的，乌泽的，表面粗糙的
matter 物质，实体，要素，成分
Matthiessen's rule 马希森（电阻率）定则
mauve 紫红色的；苯胺紫染料
maximum voltage at discharge 最大放电电压
mean time constant 平均时间常数
measly 没用的，劣质的，微小的，少量的，不充分的
measurable 可测量的，适度的，适当的
measure 量度，大小，尺寸，测量，程度，范围
measureless 无限的，非常大的，巨大的
measurement range 量程，测量范围
measuring 测量；测量的，计量的
measuring appliance 测量设备，仪表
measuring buret 量液滴定管
measuring case depth for steel 钢的表层（硬化）深度测定法

measuring column 水银柱（温度计）
measuring compressor 计测空气压缩机
measuring cylinder 量筒
measuring device 量具
measuring method 测量方法
measuring pipet 带刻度吸管
measuring scale 量〔标，刻度，比例〕尺
meaty 内容丰富的，重要的，扼要的，有力的
mechanical 机械的
mechanical chuck 机械式夹头
mechanical damage 机械损伤
mechanical impingement 机械冲击
mechanical interlocking 机械（集中）联锁
mechanical plating 机械镀
mechanical polishing 机械抛光
mechanical property 力学性能
mechanical sanding 机械打磨
mechanical spalling 机械剥落法
mechanical strength 机械强度
mechanical wear 机械磨损
mechanochemical polishing 机械化学抛光加工
medicine spoon 药匙
medium alloy steel 中合金钢
medium pressure steam 中压蒸汽
medium temperature tempering 中温回火
medium-carbon steel 中碳钢
mega 兆，百万
mega electron volt 兆电子伏特
megaphenocryst 大斑晶
megaphyric 大斑晶状的
megascope 粗视显微镜
megasonic cleaning equipment 兆频超音波洗涤设备
megger 高阻计，绝缘试验器
megavolt 兆伏特
megohmmeter 兆欧表
mekapion 电流计

melamine 蜜胺，三聚氰（酰）胺
melanin 黑（色）素
melanocratic 暗色的
meldometer （测熔点用）高温温度计
mellitic acid 苯六（羧）酸
melocol 脲-甲醛，三聚氰胺-甲醛树脂黏结剂
melt 熔化，溶解，熔体，变软
melt intercalation 熔融插层复合法
meltability 可熔性，熔度
meltable 可熔（化）的，易熔的
meltableness 可熔性，熔度
meltage 熔解量，溶解物
meltdown 熔化，熔毕，销毁
melt-growth 熔融法生长
melting 熔化
melting conditions 熔化条件
melting loss 熔损
melting point （玻璃的）熔点
melting range 熔化区域
membrane 膜
membrane curing 液膜养护
membrane filtration 膜过滤
membrane potential 膜电位
membranous 薄膜的，膜质的，膜状的
memory effect 记忆效应
mensuration 测量，测定，量度，求积法
menthol 薄荷醇（脑）
mephitical 有毒气的，有恶臭的
mephitis 毒气，恶臭
meq = milligramequivalent 毫克当量
mer weight 基体量
mercaptan 硫醇
mercerization 丝光处理，碱化，浸碱处理
mercerize 丝光处理，碱化
merchant 商业的
mercuration 汞化作用，加汞作用
mercuric 水银的，（正，二价）汞的
mercuric chloride 氯化汞，升汞
mercuride 汞化物

mercurimetric 汞液滴定的
mercurimetry 汞液滴定法
mercurization 汞化
mercurize 汞化，加汞，用汞处理
mercurous （含）水银的，（亚，一价）汞的
mercurous chloride 氯化亚汞，甘汞
mercury 汞
mercury film electrode 汞膜电极
merge 消失，吞没，熔合，溶解，汇合，合并，吸收
merge into 合并［归并，汇合］成，消失［沉没，溶解］在……之中，溶合到……里
mergence 消失，沉没，吸收，合并，结合
merging intersection 汇合交叉口
merging sort （归）并（种）类
merit 指标，准则，标准
merocrystalline 半晶质
merohedral （结晶）缺面（体）的
merohedral form 缺面形
merosymmetry （结晶）缺对称，缺面体
merotomy 分成几部分，裂成几块
mesh 筛目，筛孔，啮合，槽，孔座，罗网
mesh analysis 筛（分）析，网孔解析
mesh cathode 网状阴极
mesh filter 筛网过滤器
meshed 网状的，有孔的，啮合的
meshy 网状的，多孔的
mesial 中间的，当中的
mesokurtosis 正态峰，常态峰度
mesomeric 内消旋的，中介的
mesomeride 内消旋体
mesopore 间隙孔，中孔
meta 中（间，位），亚，元，介，偏，
meta position 间位
metabolic 变形的，同化作用的，（新陈）代谢的
metabolite 代谢物

metabond 环氧树脂类黏结剂
metachromatism （因生锈，温度变化）变色的，因光异色的
metal 金属
metal bond 金属键
metal cementation 渗金属法
metal coating 金属涂层［保护层］，包镀金属（法）
metal colouring 金属染色
metal contamination level 金属污染等级
metal electrodeposition 金属电沉积
metal elemental analysis 金属元素分析仪
metal fitting 金属配件，小五金
metal gauze 金属网
metal grill 金属格栅
metal indicator 金属指示剂
metal matrix composite (MMC) 金属基复合材料
metal penetration 金属渗透（到砂粒）间
metal plate 钣金，金属板
metal protection 金属保护
metal saw 金工用具
metal space lattice 金属结晶格子
metal spraying 金属喷镀
metal surface treatment 金属表面处理
meta-acid 偏（位）酸，间（位）酸
meta-aluminate 偏铝酸盐酸
metal-bearing 含金属的
metal-cutting 金属切削
metal-lined 有金属衬里［铺衬］的
metal-oxide 金属氧化物，金属绝缘膜
metal-partition 金属隔板
metal-plate 金属板
metal-semiconductor 金属-半导体的
metalate 使金属化
metalation 金属化作用
metalclad 金属皮，（有）金属色层的，金属包盖
metaldehyde 聚乙醛，四聚乙醛
metalepsis 取代（作用）

metaler 钣金工
metallic 金属的，含金属的
metallic bond 金属键
metallic color 金属色
metallic crystal 金属晶体
metallic enamel 金属瓷漆
metallic film 金属薄膜
metallic luster 金属光泽，金属闪光料
metallic material 金属材料
metallic soap 金属皂
metallic paint 金属漆，金属涂料
metallic seeding 金属晶种
metallic solid 金属固体
metallic sponge 海绵（状）金属
metallicity 金属性
metallics 金属粒子，金属物质
metalliferous 金属的
metallike 似金属的
metallikon 金属喷镀法，喷镀金属
metalline 金属的，含金属的，金属似的
metallization 金属化，使具有导电性；喷［敷］镀金属
metallization pattern 金属化互连图
metallize 用金属处理，使金属化
metallized 用金属处理；金属化的
metallizing 真空涂膜，金属化
metallographic examination 金相检验
metallographic 金相学的，金相的
metallography 金相学
metalloid 非金属，类［准］金属；非金属的，类似金属性的，准金属的
metalloproteinases 金属蛋白酶
metallorganic 有机金属的，金属有机物的
metalloscope 金相显微镜
metalloscopy 金属显微检查
metallostatic 金属静力学的
metallurgical 冶金的，冶金学的
metallurgical microscope 金相显微镜
metallurgical technology 金属工艺学
metallurgical thermodynamics 冶金热力学

metallurgy 冶金
metameric 位变异构的，同分异构的
metamerism 位变异构，同分异构，条件配色
metamict 蜕晶质，混胶状
metamorphism 变质，变形，变态
metamorphose （使）变化［形］
metamorphosis 变形，变质
metamorphous 变形的，变质的
metaniobate 偏铌酸盐
metaperiodic acid 偏高碘酸
metaphospate 偏磷酸盐
metaphosphoric acid 偏磷酸，二缩原磷酸
metaphrase 直译，逐字翻译；逐字翻译
metascope 红外线显示器，红外线指示器
metasilicate 硅酸盐
metastability 亚稳定性，亚稳度，亚稳性
metastable 亚稳的，相对稳定的，介稳态的
metastable austenite 介稳奥氏体
metastable ion 亚稳离子
metastable peak 亚稳峰
metathesis 置换作用，复分解作用
metathetical 复分解的，置换的
metatitanate 偏钛酸盐
metatungstate 偏钨酸盐
metawolframate 偏钨酸盐
metering device 计量仪表，测量装置，量器具
metering 测定，测量，记录，登记
methacrylate 丙烯酸酯
methacrylic acid 甲基丙烯酸，异丁烯酸
methane 甲烷，沼气
methanol 甲醇
methide （金属的）甲基化物
methodical 有系统的，有方法的
methodology 方法学，方法论
methoxide 甲醇盐，甲氧基金属
methyl 甲基
methyl ethyl ketone 甲乙酮

methyl methacrylate 甲基丙烯酸甲酯，异丁烯酸甲酯，硬［有机］玻璃
methyl orange 甲基橙
methyl phosphonate 磷酸甲酯
methyl orange 甲基橙
methylamine 甲胺
methylate 甲基化，甲醇化物
methylene 亚甲基
methylene blue 亚甲蓝
methylic 甲基的，含甲基的
methylolacetone 羟甲基丙酮
methylpentene polymer 甲基戊烯聚合物
methylphosphinate 亚膦酸甲基酯
meticulous 小心的，仔细的，精确的
metlbond 金属粘合法，（酚醛，环氧树脂类及无机）黏合剂
metol 甲氨基酚（显像剂）
metric 度量标准；公制的，米制的，公尺的
metric scale 公制尺，比例尺寸，米尺
metric size 公制尺寸
metric space 度量空间
metric unit 公制单位
metrical 测量的，度量的，韵律的，有韵律的
metrical instrument 计量仪器
metrical transitivity 度量可移性
metrically 度量上
metrication 公制化
metrizable 可度量的
metrolac 胶乳比重计
metrological 度量衡学的，计量学的
metrology 度量衡学，计量学，度量衡制，计量制
mettalic coating 金属闪光漆涂装
mezzotint 网线铜版，金属版印刷法；用网线铜版雕刻法刻
mhometer 姆欧计，电导计
miarolitic 晶洞（状），洞隙
miasma 瘴气，臭气，不良影响
micell（a） 胶束，胶态离子，微胞

micellar 胶束的，微胞的
micellar chromatography 胶束色谱法
micellar electrokinetic capillary chromatography 胶束电动毛细管色谱
micelle 胶团，胶束，微胶粒
micro 微米，微小的
micro analysis 微量分析
micro bubble 微细气泡，微泡
micro constituent 微组元，微成分
micro corrosion 显微腐蚀
micro crack 微裂
micro crystal 微晶体
micro emulsion 微乳液
micro inverse 微梯度（曲线）
micro segregation 微量离析
micro slide （显微镜的）载物片
micro structure 显微组织，微观组织
micro void filtration 微孔过滤
micro-arc oxidation 微弧氧化
micro-porous filter 微孔过滤机
microadd 微量添加
microaerophilic 微量需氧的
microalloy 微量合金
microaphanitic 显微隐晶质
microbe 细菌，微生物
microbial 微生物的，由细菌引起的
microbicidal 杀微生物的，杀菌剂的
microbicide 杀微生物剂，杀菌剂
microbiological corrosion 微生物腐蚀
microbiological 微生物的
microcallipers 千分尺，测微计
microcharacter 显微划痕硬度计
microchronometer 精密时计，测微计时表，瞬时计
microconstituent 显微组分，显微组织成分
microcopy 缩微本，由缩影胶片复印的影本，显微照片；显微照相
microcoulombmeter 微库仑计
microcrack toughening 微裂纹增韧
microcrack 显微裂纹，微观裂纹，微疵点

microcracked chromium 微裂纹铬
microdetection 微量测定
microdiecast 精密压铸
microdimensional 微尺寸的
microdispersoid 微粒分散胶体
microdot 微粒的
microeffect 微观效应，显微效应
microelectrode array 微电极阵列
microelectrode 显微电极
microelectrolytic 微量电解的
microelement 微量元素，微型组件
microencapsulation 微囊法
microetch 微刻蚀
microfarad 微法拉
microfeed 微量进给，微动送料
microfissuring 显微裂纹
microflute 微槽
microfractography 显微断谱学
microgranular 微晶粒状的
microgrid 微网［格，栅］
microhardness 显微硬度
microinch 微英寸，百万分之一英寸
microinch finishing 光制，精加工
microinhomogeneities 微观不均匀性
microlite 微晶
micromeritic 微晶粒状，微晶粒学；微晶粒状的，粉末状的
micromesh 微孔（筛）
micrometer 千分尺，测微计
micrometer caliper 千分卡尺［规］，测微器
micrometer head 测微头，千分卡头
micrometer microscope 测微显微镜
micrometer ocular 测微目镜
micrometer screw 测微螺旋
micrometer test 测微尺测试
micrometre 微米，百万分之一米
microminiaturization 微型化
micromorphology 微观形态学
microplasma oxidation 微等离子体氧化
microporous chromium 微孔铬

microporous nickel 微孔镍
microscope 显微镜
microscopic reversibility 微观可逆性
microscopic stresses 显微应力
microscopic test 金相试验
microscopy 显微学
microscopy microstructure 显微镜微观结构
microscratch 微痕
microsdjuster 微量调整器，精密调节器，精调装置
microsecond 微秒
microsection （显微）磨片，显微断面，金相切片
microsegregation 显微偏析，微观偏析，树枝状偏析
microseism 微弱的震动，脉动
microshrinkage 显微缩孔
microsize 微小尺寸，自动定寸
microsound scope 微型示波器
microspheric 微球状的
microspherulitic 微球粒状的
microstrain 微应变
microstress 微应力
microstructural 显微结构的
microstructure 显微组织，显微构造
microtest 精密实验
microthrowing power 微观分散能力
microviscometer 微型黏度计
microvoid coalescence 显微空穴聚结
microvoid 微孔
microwave 微波；微波的
microwave spectrum 微波谱
microwave inductive plasma emission spectrometer 微波等离子体发射光谱仪
microweigh 微量称量
mid-infrared absorption spectrum 中红外吸收光谱
migration 迁移
migration current 迁移电流

migratory aptitude 迁移倾向
mild air 轻微打气
mild steel plate 低碳钢板
mildew 霉；使发霉，发霉，生霉
mill 磨坊，磨粉机，压榨机；搅拌，碾磨，磨细，使乱转，被碾磨，切削
mill bed plate 底盘
mill engine 压轧机，压榨机
mill file 扁锉
mill finish 压光，滚光，挤光，精整磨轧
mill hardening 轧制余热淬火
mill limit 轧制公差
mill line 轮碾机
mill scale 黑色氧化皮
mill sheet 制造工艺规程，材料成分分析表，制造厂产品记录
mill star （清理滚筒用）三角铁，星形铁
mill stone 磨石
millability 可铣性，可轧性
millable 可轧的，适合于锯的
millbase 漆浆
milled 滚花的（表面的磨砂效果），研压的，铣成的
milled edge 铣成边
milled helicoid 铣削出的螺旋面
milled nut 周缘滚花螺母
milled ring 铣花环
milled screw 滚花头螺钉
Miller indices 米勒指数
miller 铣工
millesimal 千分之一组成的，千分之一的
milliampere 毫安
milliangstrom 毫埃
milliard 十万万，十亿
millibar 毫巴
milligram 毫克
milling-cutter 铣刀
milling-machine 铣床
mils per year (mpy) 密耳（千分之一英寸）/年

mimetic 模仿的，模拟的，类似的
mimetic crystal 拟晶
mimetic twinning 拟双晶
mimetism 模仿（性），拟态
mimic 模仿的，模拟的，假装的
mimic bus 模拟线路，模拟母线
mimic colouring 保护色
mimiced 仿制，模拟
mineral 矿物，无机物；矿物的，矿质的
mineral acid 无机酸
mineral aggregate 矿料，骨料，石料
mineral butter 凡士林，矿脂
mineral compound 无机化合物
mineral matter 矿物质
mineral oil 矿物油，液体石蜡，石油
mineral purple 氧化铁
mineral spirit 矿物油精（一种溶剂油），石油醚
mineral tar 软沥青
mineral varnish 石漆
mineral wax 地蜡
miniature 缩图，微型画，微型图画绘画术；微型的，小规模的；是……的缩影
miniaturization 小型化，微型化
minification 微小，缩小
minify 使变小，缩小尺寸，消减，贬低
minim 量滴（液量最小单位），极小的东西；最小的，微小的
minimal 最低的，最小限度的
minimal value 极小值
minimality 最小（性）
minimization 减到最小限度，估到最低额，轻视
minimize 最小化，使减到最少，使……趋于最小值，小看，极度轻视
minimum 最小值，最低限度，最小化，最小量；最小的，最低的
minimum detectable leakage rate mean free path 最小可检测泄漏率平均自由程
minimum detectable pulse width 最小可测脉冲宽
minimum detection limit (MDI) 最小检测限
minimum pulse width 最小脉冲宽
minioscilloscope 小型示波器
minitrim 微调
minium 铅丹，红铅，朱砂色，四氧化三铅
minivalence 最低化合价
miniwatt 小功率
minor 次要的，较小的，小调的，二流的
minor cycle 小周期，小循环
minor defect 次要缺陷
minor element 微量元素，痕量元素
minus charge 负电荷
minus earth 阴极接地
minus effect 副作用，不良效果
minuscule 很小的，很不重要的
minus 减，减去；负号，减号，不足，负数；减的，负的
minute 分，分钟，片刻；微小的，详细的
minute adjustment 精密调节
minute bubbles 小气泡
minute crack 发状裂隙，细裂缝
minute irregularities 微小的不平整处
miraculous 不可思议的，奇迹的
mirror reflection specula reflection 镜面反射
misalignment 不重合，未对准
misapplication 误用，滥用，非法占有
misarrange 排错，作不适当的安排，安排不当
miscalculate 算错，估计错误；算错，判断错误
miscellaneous 混杂的，各种各样的，多方面的，多才多艺的
miscellany 杂录，杂集，混杂，混合物，杂物
mischance 不幸，不幸的事，灾难，障碍，故障

mischief 伤害，损害，故障，毛病，弄坏
mischievous 有害的，有毒素的，胡闹的
mischmetal 混合稀土，稀土金属混合物，铈镧稀土合金
miscibility 可混合性，溶混性，互溶性
miscible 易混合的，可溶混的，能混溶的
miscolour 对……作歪曲叙述，颠倒黑白，把……着错色
misestimate 错估，不正确的评价
mishandle 虐待，错误地处理
mismatch 错配，不协调；使配错
misoperation 误操作，误动作
misorientation 取向错误，极向错误，错向
misoriented 使定向错误，使定位不当，使定向（定位）变异，对……指导不当
misproportion 不成比例，不匀称，不平衡，不调和
misquote 引用错误，错误引证，错误地引用
misregister 记录不准确，使（图像）重合失调（或重合不良）
misrun 浇铸不满，滞流
misshapen 畸形的，丑恶的，怪异的
misspend 浪费，滥用（时间、金钱等）
mist 薄雾，视线模糊不清，模糊不清之物；使模糊，使蒙上薄雾；下雾，变模糊
mist lubrication 油雾润滑
mist spray 喷雾润滑
mist suppressant 抑雾剂
mistermination 失谐，失配
mistiness 起雾，模糊，雾浓，朦胧
mitallation 金属化
mitigate 使缓和，使减轻；减轻，缓和下来
mitigative 缓和的，减轻的，镇静的，平静的；缓和
mitigatory 减轻的，缓和的
mitis 可锻铁
mitre 将……斜接，斜接；斜面，斜接
mix 混合，搅拌

mixable 可溶混的
mixed conductor 混合导体
mixed dislocation 混合位错
mixed indicator 混合指示剂
mixed potential 混合电位
mixed-bed exchanger 混床交换器
mixer 混合机
mixing room 混调室
mixing device vacuum mixing device 混合装置真空搅拌装置
mixture 混合物
mobile 运动物体；机动的，易变的，非固定的
mobile phase 流动相
mobility 流动性，迁移率，淌度
modal 模式的，情态的，形式的
modeling 建模
moderate 稳健的，温和的，适度的，中等的，有节制的；节制，减轻，变缓和，变弱
moderate cracking 中度裂化
moderate operating condition 中等使用［工作］条件
moderated 慢化的，适中的
moderately 适度地，中庸地，有节制地
moderator 慢化器，减速剂，缓和剂
modernize 使……现代化；现代化
moder 脉冲编码装置，中度腐殖质，酸性腐泥
modest intent 一般要求
modification 改性
modification kit 附件，改型工具
modificative 修饰的，修改的，修正的
modified 改进的，修改的，改良的
modified austempering 变质等温淬火
modified constant voltage charge 修正恒定电压充电
modified index 修正指数
modifier 改性剂，改良剂
modifing agent 改良剂，变换剂，改善剂
modify 修订，更改，变更

modulated 已调的，被调的
modulated current plating 调制电流电镀
modulation analysis 调制分析
modulatory 调节的，调制的
modulus 模数
modulus of elasticity 弹性模量
modulus of resilience 回弹系数，回能模数
modulus of rigidity 刚性模量
modulus of rupture 挠曲强度，极限强度，折断系数，裂断模量，断裂模数
modulus of torsion 扭转模量
Mohs' hardness 莫士硬［度］标
moiety 一部分，一半，二分之一
moist 潮湿的，多雨的；潮湿
moisture apparatus 测湿器
moisture 水分，湿度，潮湿，降雨量
moisture capacity 湿度
moisture content 含水量
moisture eliminator 去潮器，干燥器，脱湿器
moisture expansion 水分膨胀
moisture film 湿膜
moisture permeability 透湿性，透水汽性
moisture regain 吸湿（性），回潮率，回潮
moisture retention 保水性
moisture scavenger 去湿剂
moistureless 无湿气（或水分）的，干燥的
moistureproof 防潮的；使防潮湿
moisturize 使增加水分，使湿润；增加水分，变潮湿
molality 质量摩尔浓度（每1000克溶剂中溶质的物质的量）
molar 质量的，摩尔的
molar absorptivity 摩尔吸光系数
molar concentration 物质的量浓度，容模浓度
molar fraction 体积克分子分数，摩尔分数
molar mass 摩尔质量

molar weight 摩尔量，分子量
molarity 摩尔浓度
mold 模具，塑造，使发霉，用模子制作
mold core 模芯，型芯
mold polishing 模具打磨
mold release 脱模
molding 成型，模塑
molding equipment 塑模成型装置，封胶装置
molding press 封胶冲压
molding time 成型时间
molecular 分子的，由分子组成的
molecular chemistry (polymer) 分子式（聚合物）
molecular fluorometry 分子荧光分析法
molecular forces 分子间力
molecular geometries 分子空间构型
molecular ion 分子离子
molecular orbital theory 分子轨道理论
molecular weight 分子量
molecule bond 分子键
molybdate 钼酸盐
mole 摩尔
molecule 分子，微小颗粒
moly high speed steel 钼高速的钢
molybdenic （三价）钼的
molybdenous （二价）钼的
molybdenum 钼
molybdenum steel 钼钢
molybdic （正，三价，六价）钼的
molybdous 亚（二价）钼的
momentary discharge 瞬间放电
monacid 一元酸
monadical 一元的
monamide 一酰胺
monamine 一元胺
monatomic 单原子的，单质的，一价的
monatomic acid 一元酸
monistical （溶液中）的未电离（未游离）的
monitor 监视器

monitored 检测
monitoring 监测
mono molecule layer adsorption 单分子层吸附理论
monoblock 整体的；整体，直板，单块
monoblock container 整体槽
monochromator 单色器，单色光镜
monolayer 单层，单层的
monolithic 单块集成电路，单片电路；整体的，巨石的，庞大的，完全统一的
monomer 单体，单元结构
monomeric 单体的，单节的，单分子构造的
monomeric unit 链节，单体单元
mono- 【构词成分】单的，单一的，一
morphology 形态学，形态论，形态
mosaic 马赛克，镶嵌，镶嵌细工；拼成的，嵌花式的
mother liquor 母液
mottle 斑点，杂色，斑驳；使呈杂色，使显得斑驳陆离
mottling 斑点，麻点
mould release agent 脱模剂，下模剂
mould split line 模具分型线
mouldability 可塑性
moving 移动的，活动的
moving bed 移动床
muffle furnace 马弗炉，回热炉
multi-bath wet cleaning equipment 多槽浸渍式洗涤装置
multi-chamber vacuum system 多室真空系统
multi-cycle annealing 多循环退火处理
multi-frequency 多频
multi-site probing 多部位探测
multi-station cleaning equipment 多处理站洗涤装置
multi-station synchronous probe test 多工位同步探测试验
multi-step annealing 多步骤退火处理
multi-step fourier transform infrared spec-troscopy (MSFTIRS) 多步骤电势阶跃傅里叶变换红外光谱学
multicoloured 多彩的
multicomponent analysis 多组分分析
multilayer 多层
multilayer plating 多层电镀
multimeter 万用表
multimode electrochemical detection 多模式电化学测量
multiplayer resist method 多层抗蚀剂法
multiple back reflections 多次底面反射 多次回波法
multiple bonds 重键
multiple interference effect 多重干涉效应
multiplex 多样的，多元的，多重通道
multiplicate 多种的，多重的，并联的
multiplication 乘法，增加，放大，按比例增加
multiplicative 倍增的，乘法的，增殖的
multiplicity 多样性，多重性，重复性
multiplier 倍增器，增加者，繁殖者
multiply 多层的，多样的；乘，使增加，使繁殖，使相乘；并联地，多样地，复杂地
multiply connected 多连通的
multiply periodic 多周期的
multiplying factor 倍率，放大率
multipole 多极，复极；多极的
multipole switch 多极开关
multipolymer 共聚物
multiprobe 多功能探针
multipurpose pliers 万能手钳
multiscale analysis 多尺度细化分析
multistage 多级的，多阶段的，多节的；使分成多阶段
multistep 多级的
multivariant 多元，多自由度的，多方案的，多变的
municipal water 城市用水
muriate 氯化物
muriatic 氯化的，盐酸化的

muriatic acid 工业盐酸

murk 黑暗，阴沉，阴郁；阴郁的，黑暗的

mush 烂泥，软块，糊状物

muslin 棉布，平纹细布

mustard 芥末，芥子气

mustard oil 芥子油，异硫氰酸酯

mutagen 诱导有机体突变的物质，致变物

mutamer 变构物，旋光异构体

mutamerism 变旋光现象，变构现象

mutarotation 旋光改变

mute 沉默的，无声的；减弱……的声音；使……柔和

mutilate 切断，毁坏，使……残缺不全，使……支离破碎

mutilative 破坏性的，切断的

mutual 共同［相互，彼此］的

mutual attraction 相互吸引

mutual effect 相互作用

mutual solubility 互溶性

mutuality 相互关系，相关，亲密

mutually 互相地，互助

mutually exclusive 互不相交的，互斥的

myriad 无数，极大数量，无数的人或物；无数的，种种的

myriametric wave 超长波，一万公尺，频率小于 3 万赫兹

myriametric 万公尺的

myrmekitic structure 蠕状构造，蠕状结构

mythological 凭空想象的，虚构的

N

nadel 针状突起
nadir 最低点，最底点
nail 指甲，钉子；钉，使固定，揭露
naked eye 肉眼
naked radiator 无保护罩散热器
naked wire 裸线
naked 裸露的，无装饰的，无保护的，无证据的，直率的
nalcite 离子交换树脂
named 命名的，指定的，被指名的
nano- 【构词成分】纳，毫微
nanocarbon tube 碳纳米管
nanocrystalline ceramics 纳米晶陶瓷
nanofiltration 纳滤
nanolithography 纳米刻蚀，毫微光刻
nanoparticles 纳米粒
naphthalene 卫生球，臭樟脑
naphtha 石脑油，挥发油，粗汽油
naphthene base 环烷（烃）基
naphthene hydrocarbons 环烷（烃）基
naphthene 环烷，环烷（属）烃
naphthenic acid 环烷酸，环酸
naphthenic 环烷的，（脂）环烃的
naphthenone 环烷酮
naphthol 萘酚
naphthylamine 萘胺，甲萘胺
nascence 起源，发生
nascent hydrogen 初生氢，新生氢
nascent state 初生态，新生态
nascent 初期的，开始存在的，发生中的
nastily 污秽地，不洁地，讨厌地
nasty 令人不快的事物，下流的，肮脏的，难以应付的
national electrical code 国家电气规程
native asphalt 天然沥青
native oxide layer 自然氧化膜
native 本国的，土著的，天然的，与生俱来的，天赋的
natural abrasive 天然研磨料
natural aging 自然时效
natural crack 自然裂纹
natural frequency 固有频率
natural gas 天然气
natural oxidation 自然氧化
natural parameter 特征参数
natural pattern 实物模
natural resin (oil) 天然树脂（油）
natural 自然的，物质的，天生的，不做作的
naturally 自然地，必然地，容易地
nature 自然，性质，本性，种类
navigation 航行，航海
nay ionic 非离子
near surface defect 近表面缺陷
near ultra-violet rays 近紫外线
near-critical 近临界的
nearside 靠近的一边；靠人行道的，左侧的
near-spherical 近似球形的，类球状
neatline 图表边线，图廓线，准线
neatsoap 净皂，纯皂
neat 灵巧的，整洁的，优雅的，齐整的，未掺水的，平滑的
nebular 星云（状）的
nebulization 雾化，气化
nebulize 喷洒，使……成雾状，使……成喷雾状
nebulosity 朦胧，星云状态，星云状物
nebulous 朦胧的，星云的，星云状的
necked-in 向内弯曲
necked-out 向外弯曲
necked 收缩的，变窄的
necking 缩颈
needle 针，指针，刺激，针状物；刺激，用针缝，缝纫，做针线

needle-like 针状的
negate 对立面，反面；否定，取消，使无效，否定，否认，无效
negation 否定，否认，拒绝
negative electrode 负极
negative leveling 负水准
negative resistance 负电阻
negative 否定，负数，底片；负的，消极的，否定的，阴性的；否定，拒绝
negligence 疏忽，忽视，粗心大意
negligible 可忽略
negotiability 流通性，可转让性，可磋商性
negotiable 可通过谈判解决的，可协商的
negotiate 谈判，交涉；谈判，商议，转让，越过
negotiation 谈判，转让，顺利的通过
negotiatory 商议的
nematic 向列的
neodymia 氧化钕
neogenesis 新生，再生
neogenic 新生的
neohexane 新己烷
neomagnal 铝镁锌耐蚀合金
neomorphic 新生形
neon 霓虹灯，氖
neopentane 新戊烷
neoprene 氯丁橡胶
neosome 新成体
neoteric 现代的，新发明的，新颖的
neozoic 新生代的
nepheloid layer 雾状层
nephelometer 浊度计，悬液计，比浊计
nephelometry 用悬液计测量悬液，散射测浑法，浊度测定法
Nernst equation 能斯特方程
nesslerization 等浓比色法
nest 嵌套，群，组，束，巢穴；把……套起来
net weight 净重
net work 净功

nethermost 最下面的，最低的
netro compound 硝基化合物
net-shaped 网状的
netted texture 网状结构
netted 网状的，用网包［捕］的
netting 网，网鱼，结网
netty 网状的
network polymer 网状聚合物
network 网状物，网络，广播网
net 网，网络，净利，实价；纯粹的，净余的
neuter 中性的，不及物的，无性的
neutral carrier 中性载体
neutral salt spray test (NSS-test) 中性盐雾试验
neutralization tank 中和池
neutralization 中和
neutralizer 中和剂，中和器
neutralize 抵消，使……中和，使……无效，使……中立，中和，中立化，变无效
neutralizing 中和，平衡，抵消
neutral 中和的，中性的，中间的，不带电的，非彩色的
neutretto 中性介子
neutron radiography 中子射线照相术
neutron 中子
newborn 新生的，再生的
newfashioned 新型的，新式的，新流行的
niacin 烟酸，尼克酸
nib 尖端，尖头，尖劈，尖楔，突边
nibble 轻咬，啃，细咬，一点一点切下
nibbling 步冲轮廓法，分段剪切
nicarbing 碳氮共渗，气体表面硬化法，气体（氰化）
Nichicon 电容器
nichrome 镍铬铁合金，镍铬耐热合金
nick action 交咬作用
nick bend test 刻槽挠［弯］曲试验，缺口弯曲试验
nick 划痕
nickel bath purifier 镍浴除杂剂

nickel cadmium batteries　镍镉电池
nickel carbonyl　羰基镍
nickel chloride　氯化镍
nickel dipping　镍盐浸
nickel iron batteries　镍铁电池
nickel metal hydride batteries　金属氧化物镍氢电池，镍氢电池
nickel sealing　镍盐封孔
nickel silver　镍黄铜，德银
nickel steel　镍钢
nickel strike　闪镀镍
nickel sulphamate　氨基磺酸镍
nickel sulphate　硫酸镍
nickel zinc batteries　镍锌电池
nickelage　镀镍
nickelous　二价镍的，亚镍的
nickel　镍；镀镍于
Nicol　尼科尔棱镜（偏光镜）
nicotine　尼古丁，烟碱
nicotinic　烟碱的，烟碱酸的
niggerhead　黑礁砾，低劣的橡胶，不熔块
nigrescence　变黑；（眼睛、头发、皮肤等的）黑色
nigrescent　发黑的，带黑的，易于变黑的
nigre　皂脚
nigrify　使……成黑色，变黑
nigrometer　黑度计
nilometer　水位计
nil　无，零
nimiety　过多，过剩
Nimol　尼莫尔铁，尼莫尔耐蚀高镍铸铁
niobate　铌酸盐
niobic　（五价）铌的，含铌的
niobium　铌
niobous　三价铌的，亚铌的
nipper　镊子
nipple　螺纹接头
nip　夹，捏，掐，挤，咬，压缩，剪断，虎钳
nital　硝酸乙醇浸蚀液
nitralising　硝酸钠溶液浸渍净化法

nitralloy　氮化合金钢
nitrate　硝酸盐；用硝酸处理
nitration　硝化，用硝酸处理，硝基置换
nitre　硝酸钾，硝石
nitriability　氮化性
nitric acid　硝酸
nitric oxide　氧化一氮
nitric　氮的，含氮的，硝石的
nitridation　氮化，渗氮
nitride　氮化物，渗氮，硝化
nitriding　氮化，渗氮
nitrification　硝化
nitrify　硝化，使与氮化合，用氮饱和
nitrile　腈，腈类
nitrilotriacetic acid（NTA）　氨三乙酸
nitrite　亚硝酸盐
nitrizing　氮化法；渗氮
nitroaniline　硝基苯胺
nitrobenzene　硝基苯
nitrocarburizing　软氮化，氮碳共渗
nitrocellulose lacquer　硝基漆
nitrocellulose　硝化纤维
nitrocompound　硝基化合物
nitro-derivative　硝基衍生物
nitrogen case hardening　渗氮，氮化
nitrogen dioxide　二氧化氮
nitrogen group　氮族
nitrogen group element　氮族元素
nitrogen hardening　渗氮硬化
nitrogenation　氮化作用
nitrogen-free　无氮的
nitrogenous　含氮的
nitrogen-sealed　氮气密封
nitrogen　氮
nitroso compound　亚硝基化合物
nitroso　亚硝基的
nitrosyl　亚硝酰基；亚硝酰基的
nitrous　氮的，硝石的，含氮的
nitroxyl　硝酰（基）
nitrozation reaction　亚硝基化反应
nitrozation titration　亚硝基化滴定法

nitrozation 亚硝化作用
nitro 硝基
NMR spectroscopy 核磁共振波谱法
NMR spectrum 核磁共振波谱
nobility 贵金属性
noble metal 贵金属
noble 惰性的，不易起化学作用的
noctilucence 生物发光，夜间发光
noctilucent 夜光的，生物发光的，夜间可见的
nodal 节的，结的，节似的
node 节点，瘤，叉点
nodical 交点的
nodular cast iron 球墨铸铁
nodular cementite 粒状渗碳体
nodular graphite 球状石墨
nodular 结节状的，有结节的
nodulation 结瘤
nodulizing 球化，附聚作用，烧结作用
noise resistant 噪声抗
noise 噪声
noisome 恶臭的，有害的
nominal voltage 标称电压，额定电压
nominal 标称的，标额的，极小的
nomogram 诺模图，列线图，计算图表
nomographic chart 列线图，算图
non oriented 非定向，非取向的
non periodic 非定时的，非周期性的
nonactivated 未活化的，未激活的
nonaging 不老化的，不陈化的，经久的，无时效
nonanedioic acid 壬二酸
nonane 壬烷
nonanol 壬醇
non-aqueous dispersion paint 非水分散型涂料
non-aqueous liquid developer 非水性液体显像剂
non-aqueous solvent 非水溶剂
nonaqueous titration 非水滴定法
non-aqueous 非水的
non-automatic 非自动的
non-axial 非轴（向）的
non-caking 不结块的
non-coherent 非相干性的，无黏聚力的，不附着的
noncombustible 不燃物；不燃的
non-commutative 非交换；非交换的
non-condensable 不凝的
non-conducting 不导电的，不传导
nonconductor 绝缘体，非导体
nonconservation 不守恒
nonconsumable 非自耗的
noncoplanar 非共面的
non-corrodible 防腐的，抗腐蚀的
non-corrosiveness 无腐蚀性
non-corrosive 无腐蚀性的，不锈的
non-countable 不可数的
non-criticality 非临界性
noncrystalline electrodes 非晶体电极
noncrystalline 非晶态的，非结晶的
noncubic 非立方系的
non-cyanide copper plating 无氰镀铜
noncyclic 非旋回的，非周期的
nondeflecting 不变形的，不挠曲的
nondeformable 不变形的
nondense 疏的，无处稠密的
nondestructive inspection（NDI） 无损检验，非破坏性检验
nondestructive tests 非破坏性测试，无损测试
nondestructive 无损的，非破坏性的
nondrying oil 不干性油
non-elasticity 非弹性，无伸缩性
nonentity 不存在（的东西），虚构（物）
non-equilibrium 非平衡，不均衡
nonequivalent 非等值的，非等效的
non-faradaic current 非法拉第电流
nonfaradaic processe 非法拉第过程
nonferromugnetic material 非铁磁性材料
nonferrous alloy 非铁合金
nonferrous metal 有色金属

nonfissionable 不可分裂的
nonflammable 不燃烧的，不易燃烧的
non-flowing 不流动的
non-fluctuating 非脉动的
non-holonomic 非完整的
non-homogeneity 不均匀性，非均质性
non-inductive 非诱导的；无电感的
non-inert impurity 非惰性杂质
non-inflammability 不燃性，具不燃性
noninjurious 无害的
non-interacting 非相互作用的，不互相影响的
non-interference 不干涉，不干预
non-ionic surface active agent 非离子表面活性剂
nonionic surfactant 非离子表面活性剂
non-ionic 非离子的
non-isothermal 非等温的
nonisotropic 各向异性的，非各向同性的
nonlinearity 非线性，非线性特征
nonlinear 非线性的
non-loaded 无负载的
nonlocalizability 不可定位性
non-mechanized 非机械化的
nonmetallic coating 非金属涂层
non-metallic inclusion 非金属夹杂物
nonmetallic 非金属的
nonmetal 非金属
non-miscible 不互溶的
nonmutilative 非破坏性的
nonnegligible 不可忽视的（重大的）
non-normality 非常态性，不垂直，非正态性
non-orthogonal 非正交的，非正角正交
nonorthogonality 非正交性
nonoverflow 非溢流
non-oxidizability 不可氧化性
nonpolar group 非极性基团
nonpolar molecule 非极性分子
nonpolar 非极性的，无极的
nonpolarity 非极性

non-polishing 不易磨光
non-porous 无孔的
nonreactivity 惰性，无反应性
nonreciprocal 单向的，非交互的，非互易的
nonrecoverable 不能收回的，不可恢复的
nonrectification 不能整流
nonreflecting 不反射的
non-return valve 逆止阀，单向阀
nonrigid 非刚性的
non-rusting 不锈的，防锈的
non-saponifying 不皂化的，不可皂化的
nonsaturable 不饱和的
non-selective 非选择性的
nonselfignition 非自燃的
nonsettling 不沉降的
non-shattering 不易脆的
non-shrinking 不收缩的
non-skid 防滑的
nonslaking 不水解的
nonslip 防滑的
nonsoluble 不溶解的
non-solute 非溶质，不溶质
nonstationarity 非平稳，非恒定
non-stationary 不稳定的，非平稳的
nonsteady state diffusion 非稳态扩散
nonsteady 不稳定的
nonsticking 不黏附的
nonstoichiometry 非化学计量性（偏离化学计量比），非定比性，非理想配比性
non-swelling 不膨胀的
non-symmetrical 非对称的
non-synchronous 非同步的，异步的
non-toxic 无毒性；无毒的
non-transition metal 非过渡金属
non-transparency 不透明度，不透明性
non-uniform 不均匀，非均一
nonuniformity 不均匀性，非一致性
nonvariant 不变的
non-viscous 无黏性的，不黏的
nonvolatile content 不挥发分

nonvolatile 非易失的，不挥发的
non-volatility 不挥发性
non-wettable 不可湿润的，不可润湿的
non-wetting 焊不良，不润湿，不沾锡
nonwoven separator 非织造分离器
non 非，不，无
normal capacity 正常容量，额定容量，正常生产能力
normal cross section 横截面，标准剖面，正剖面
normal distribution 常态分配，正态分布
normal hydrogen elect rode（NHE） 标准氢电极
normal hydrogen electrode（NHE） 标准氢电极
normal permeability 标准磁导率
normal pulse polarography（NPP） 常规脉冲极谱法
normal pulse voltammetry（NPV） 常规脉冲伏安法
normal pulse voltammetry 常规脉冲伏安法
normal pulse 常规脉冲，正常脉冲
normal solution 当量溶液，规度溶液
normal voltage 正常电压
normal wear 正常磨损
normal 正常，标准，常态；正常的，正规的，标准的
normality 常态，标准状态，正规性，当量浓度，规定浓度
normalization method 归一化法
normalize 使正常化，使规格化，使标准化
normalized reactance 归一化电抗，标准化电抗，标准化电阻
normalizing 正火，正常化；使正常化，使正规化，对钢正火
normative 规范的，标准的
normteile 标准件

notability 显著，著名，值得注意
notate 以符号表示
notation 符号，乐谱，注释，记号法
notchboard 梯级搁板，凹板
notched 有凹口的，有缺口的，有锯齿状的
notching 下凹的，多级的；冲口加工；刻凹痕，用刻痕计算
notch 刻痕，凹口，等级，峡谷；赢得，用刻痕计算，在……上刻凹痕
notedly 显著地，知名地
notional 概念性的，想象的，抽象的，不切实际的
no-touch 不接触，无触点
nought 零，没有
noumenal 本体的，实体的
noumenon 本体，实体
novolac （线型）酚醛清漆
noxious 有害的，有毒的
nozzle velocity 喷嘴速度
nozzle 喷头，喷嘴
nubbly 块状的，多瘤的
nucleal 核的
nuclear magnetic resonance（NMR） 核磁共振
nucleate 有核的；使成核，成核
nucleation 形核现象，晶核形成
nuclide 核素
nugatory 无价值的，琐碎的，无效的
nullification 无效，废弃，取消
nullify 使无效，作废，取消
null 零，空；无效的，无价值的，等于零的
nulvalent 零价，不活泼的，不起反应的
nutsch 吸滤器
nut 螺母，螺帽，难对付的人，难解的问题
Nykrom 高强度低镍铬合金钢
nylon 尼龙

O

Oberhoffer solution 钢铁显微分析用腐蚀液
object beam angle 物体光束角，物体波束角
object glass 物镜
object plane resolution 物体平面分辨率
object scattered neutrons 物体散射中子
objectify 使具体化，使客观化，体现
objection 异议，反对，缺陷，缺点，妨碍，拒绝的理由
objectionable 讨厌的，会引起反对的，有异议的
objective 目的，目标，物镜；客观的，目标的
object-line 轮廓线
object-staff 准尺
oblate 扁圆的，扁球状的
obligate 使负义务，强使，强迫，对……施以恩惠；有责任的，有义务的，必需的
obligate aerobes 专性需氧微生物，专性需氧菌
obligation 义务，职责，债务
obligatory 义务的，必须的，强制性的
oblique 倾斜物；斜的
obliqueness 倾斜，斜度，歪斜
obliterate 消灭，涂去，冲刷，忘掉
oblong 椭圆形，长方形；椭圆形的，长方形的
obscuration 遮蔽，昏暗，晦涩
obsure 指不清晰，模糊
observability 可观察性，能观测性
observable 可观察量，感觉到的事物；显著的，觉察得到的，看得见的
observation 观察，监视，观察报告
observing 观察的，注意的，观察力敏锐的
obsolete 废弃的，老式的
obsoletism 废弃

obstacle 障碍，干扰，妨害物
obstruct 妨碍，阻塞，遮断；阻塞，设障碍
obstruction 障碍，阻碍，妨碍
obstructive 妨碍物，障碍物；阻碍的，妨碍的
obturate 封闭，填塞，密闭
obtuse 迟钝的，圆头的，不锋利的
obverse 正面，正面的
obviate 排除，避免，消除
occlude 使闭塞，封闭，挡住，咬合
occluded water 包埋（藏）水
occult 神秘的，超自然的，难以理解的；掩蔽，被掩蔽
occupancy 居住，占有，占用
octad 八价元素
octadecane 十八烷，正十八烷
octadecene 十八烯
octadecyl 十八烷基
octadic 八价的
octahedral 八面体的，有八面的
octahedral compound 八面体化合物
octahedral position 八面体配位
octahedrite 锐钛矿（八面石，八角形二氧化钛晶体）
octahedron 八面体
octamer 八聚物
octane 辛烷
octanol 辛醇
octavalence 八价
octyl 辛基
octylene 辛烯
odd 奇数的，剩余的，临时的，零散的
odor 气味
odorants 有气味的东西；有气味的，有香气的
odorous 香的，有气味的，难闻的
odourless 无气味的，无臭的

oeolotropic 各向异性的
offal 垃圾，碎屑，工业下脚
off color 色差，色泽不佳的
off-gas 尾气，出口气，废气
offgrade 不合格的，品质低劣的
off-load voltage 空载电压，开路电压
offscum 废渣
offset 抵消，补偿
offsetting 位移，位移指线路，斜率，不均匀性
ogee 弯曲，双弯曲线，S形；双弯曲线的，S形的
ogival 尖顶式的；尖拱的，蛋形的，卵形的
Ohm's law 欧姆定律
Ohmic curve 欧姆曲线，电阻曲线
Ohmic drop 欧姆压降
oil absorption value 吸油量
oil cratering 油缩孔
oil free alkyd resin 无油醇酸树脂
oil grease 油脂
oil paint 油性涂料
oil quenching 油淬火
oil resistance 耐油性
oil separtor 隔油池
oil soluble dye 油溶性染料
oil stains 油污，油迹
oil-fired furnace 油炉
olamine 乙醇胺
oleaginous 油质的，油腻的
oleamide 油酸酰胺
oleate 油酸盐，油酸酯
olefine 链烯烃
olefinic 烯族的
oleic acid 油酸，十八烯酸
oleiferous 产油的，含油的
olein 三油酸甘油酯，油精，脂肪的液状部分
oleosol 润滑油
oligomerization 低聚，齐聚反应

omegatron 回旋质谱仪，高频质谱仪
omnibus 综合性的，总括的，（包括）多项的
omnibus bar 汇流条，汇流排，母线
omnidirectional 全方向的，无定向的
ondograph 高频示波器
on-hand inventory 现有库存
on-line analysis 在线分析
on-load voltage 负载电压
onset 开始，着手
opacifier 乳白剂，遮光剂，不透明剂
opacity 不透明，不透明性，蔽光性，暗度
opal 乳白的
opalescence 蛋白色光，乳白光
opaque 不透明物；不透明的，不传热的；使不透明，使不反光
open chain compound 开链族化合物
open-circuit 开路的
open circuit potential 开路电位
open circuit voltage 开路电压
open setting 松装
openwork 网状细工，露天采掘场；有网状小孔的
operability 可操作性
operating cycle 运行周期
operation 施工，操作
operation procedure 作业流程，操作规程
opportune 适当的，恰好的，合时宜的
opposite 相反的；在……的对面
oppositely 反向地，在相反的位置，面对面
optical 光学的，视觉的
optical activity 旋光性
optical memory materials 光存储材料
optical metallographic examination 光学金相检验
optical microscopy 光学显微镜
optical property 光学特性，光学性质
optical property tester 光学性能测定仪
optical waveguide fiber 光导纤维

optically transparent electrode (OTE) 光透明电极，光透电极
optically transparent thin-layer electrode 光透明薄层电极
opticity 光偏振性，旋光性
optimal 最佳的，最理想的
optimization 最佳化，最优化
optimum 最佳效果，最适宜条件；最适宜的
optimum point of coagulation 最优凝聚点，混凝最佳点
option 选项，选择
optional 可选择的，随意的
orange peel 橘皮状表面缺陷
orb 球，天体，圆形物；成球形，弄圆，围
orbed 圆形的，球状的，十全的
orbicular 圆的，球形的，完整的，环状的
orbital 轨道
orbital overlap 轨道重叠
ordered 有序的，整齐的，安排好的
ordinal 顺序的，依次的
ordinance 条例，法令，布告
ordinarily 通常地，一般地
ordination 分类，整顿，排列，命令
organic 有机的，组织的
organic additives 有机添加剂
organic binder 有机接合剂
organic chemistry 有机化学
organic (color) pigment 有机（颜色）颜料
organic material 有机材料
organic pigment 有机颜料
organic silicon paint 有机硅漆
organic solvent 有机溶剂
organic solvent degreasing 有机溶剂脱脂
organolite 离子交换树脂
organonitrogen 有机氮
organophosphorus 有机磷的
organosilane 有机硅烷
organosilicon 有机硅
organosol 有机溶胶，增塑溶胶
orient 东方的；使适应，确定方向，定向，定位
orientation 方向，定向，适应
oriented film 取向薄膜，定向膜
orifice 孔口，小孔，小洞
original 原始的，最初的，独创的，新颖的
ornament 装饰，修饰
ornamentally 装饰地，用作装饰品地
ortho position 邻位
orthoester 原酸酯
orthogonal 正交直线；正交的，直角的
orthogonality 正交性，相互垂直
orthograph 正投影图
orthographic drawing 正投影图
orthographic projection 正投影
orthohexagonal 正六方形的，正六方的
orthonormal 标准正交的，正规化的
orthonormality 正规化
orthophosphate 正磷酸盐
orthosilicate 正硅酸盐
oscillate 使振荡，使振动，使动摇
oscillation 振荡，振动，摆动
oscillator 振荡器
oscillograph 示波器，记录仪，波形图
oscillometer 示波计
oscillopolarographic titration 示波极谱滴定
oscillopolarography 示波极谱法
oscilloscope 示波器，示波镜
osmometer 渗压计
osmometry 渗透压力测定法
osmondite 奥氏体变态体
osmosis 渗透，渗透性，渗透作用
osmosize 渗透
osmotaxis 趋渗性
osmotic 渗透性的，渗透的
osmotic coefficient 渗透系数
osmotic pressure 渗透压
ostensible 表面的，假装的

ostensibly 表面上，外表
ostensive 清晰显示的，以实例证示的
outage 中断供应，运行中断
outdoor exposure test 屋外曝晒试验，自然曝晒试验
outer helmholtz plane 外亥姆霍兹面
outfit 机构，用具，全套装备
outflow 流出，流出量，流出物
outlet 出口，排放孔
outlet water 废水，排水
outline 外形线；轮廓线
outset 开始，开端
outside heating method 外部加热法
outside indicator 外指示剂
outstretch 拉长，伸展得超出……的范围
outstrip 超过，胜过
outward 向外的，外面的；外表，外面；向外，在外
outwear 比……经久耐用，用旧
outworn 陈腐的，用旧的，疲惫的
oval 椭圆形，卵形；椭圆的，卵形的
ovality 椭圆，椭圆度，卵形
ovaloid 卵形面；卵形的
oven 炉，灶，烤炉
over charge 过充，超载
over coating 重叠涂装，过量涂层
over development 显影过度，过度显影
over discharge 过放电
over emulsification 过度乳化，乳化过度
over etching 过腐蚀
over flow rinse 溢流冲洗
over pickling 过浸渍
over potential 超电势，过电位
over spray 过喷涂
over travel 超程，多余行程
overabound 过量存在，过于充足（或富足）
overabundance 过多，过于丰富
overage 过多的，过剩的，过于老化的
overaging 过老化，过时效
overall 全部的，全体的，一切在内的；全部地，总的说来
overall magnetization 整体磁化
overall stability constant 总稳定常数
overbake resistance 耐过烘性
overbleach 过漂
overbrim 使满出；溢出，满出
overcapacity 生产能力过剩，超负荷
overcharge 过度充电，超载
overcharge life test 过充电寿命试验
overcoatability 面漆配套性，再涂性
overcompensate 给予……过分的补偿，过度补偿
overcompression 过压缩
overcool 过冷，过度冷却
overcrowded 过度拥挤的，塞得太满的
overdose 药量过多，过度剂量
overdraft 透支，过度通风
overdried 使太干，使过干
overdye 把……染得过深，把……染得过久，套染
overexpansion 过度膨胀
overfall 溢流，湍流
overfeed 过装料，过量进料
overfill 把……装得溢出，装得太多
overflow 使溢出，充满
overflowing 溢流，过剩，溢出物；过剩的，溢出的，充满的
overflow pipe 下导管，溢流管
overfrequency 超频率，过频率，超过频率
overgild 给……表面镀金，把……染成金色
overgrind 研磨过度；过度粉碎
overgrown 蔓生的，生长过快的
overhang 悬于……之上，伸出在……之外
overhardening 过硬的
overhaul 彻底检修，详细检查；分解检查，大修
overhaul life 大修周期
overheating 过热
overlade 使超载，使装载过多
overlap 重叠，重复；与……重叠，与

……同时发生
overlay 在表面上铺一薄层，镀；覆盖图，覆盖物
over-neutralization 过度中和
overpotential 过电位，超电势
overquench 过冷淬火，淬火过度
overrange 超出额定界线的，过量程的
overrate 估计过高，高估，超过额定值
overrelaxation 过度松弛
overriding 占优势的，首要的，扼要的
overshadow 使失色，使阴暗，遮阴，夺去……的光彩
overspill 溢出物
overstrain 过度应变，紧张过度，超载
overstretch 过度伸长，过拉伸
overt 明显的，外表的
overtemper 过度回火
overtension 过应力，超限应力，电压过高
ovoid 卵形的，圆形的
oxa- 【构词成分】氧杂，噁
oxalate 草酸盐，乙二酸盐
oxalic acid 草酸，乙二酸
oxalic acid anodic oxide coating 草酸阳极氧化膜
oxidability 可氧化性
oxidable （可）氧化的
oxidant 氧化剂
oxidant inhibitor 防氧化剂
oxidate 氧化，氧化物
oxidation 氧化（作用）
oxidation aging 氧化老化
oxidation and decarbonization 氧化脱碳
oxidation current 氧化电流
oxidation number 氧化数
oxidation fog 氧化灰雾
oxidation number 氧化数，氧化值
oxidation reduction potential 氧化还原电位
oxidation-reduction reactions 氧化还原反应

oxidation-reduction titration 氧化还原滴定法
oxidation wave 氧化波
oxidative decarboxylation 氧化脱羧
oxide 氧化物
oxide cell 氧化膜单元
oxidic 氧化的
oxidiferous 含氧化物的
oxidimetry 氧化还原滴定
oxidizability 氧化性能，易氧化度，可氧化性
oxidizable 可氧化的
oxidization 氧化，氧化作用
oxidize 使氧化，使生锈；氧化
oxidized 被氧化的
oxidizer 氧化剂
oxidizing agent 氧化剂
oxidizing flame 氧化焰
oxido- 氧化
oxime 亚硝基化合物，肟
oxonium 氧鎓
oxonium ion 水合氢离子
oxo-process 氧化法，氧化合成，羰基合成
oxo-synthesis 羰基合成
oxyacetylene 氧乙炔；氧乙炔的
oxyacid 含氧酸
oxyamination 羟氨基化
oxychloride 氯氧化物
oxydol 双氧水，过氧化氢
oxydrolysis 氧化水解
oxyferrite 氧化铁素体
oxyful 过氧化氢，双氧水，变氧水
oxygen 氧气
oxygen concentration monitor 氧气浓度监控器
oxygen concentration cell 氧浓差电池
oxygen consuming (OC) 耗氧量
oxygen group 氧族元素
oxygen liberation 氧气释放
oxygen wave 氧波

oxygenant 氧化剂
oxygenate 氧化，充氧，以氧处理，使……与氧化合
oxygenic 氧的，含氧的，似氧的
oxygenolysis 氧化分解作用
oxyhalide 卤氧化物
oxyhydrate 氢氧化物
oxymercuration 羟汞化
oxymuriate 氯酸盐
oxynitrate 含氧硝酸盐
oxynitride 氮氧化物
oxysulfide 硫氧化物，含氧硫酸盐

ozocerite 地蜡，石蜡
ozonation 臭氧处理
ozonator 臭氧发生器
ozone 臭氧，新鲜的空气
ozone resistance test 抗臭氧试验
ozonic 臭氧的，含臭氧的，臭氧般的
ozonide 臭氧化物
ozoniferous 含臭氧的，产臭氧的
ozonized ultrapure water 臭氧化超纯水
ozonizer 臭氧发生器，臭氧管
ozonolysis 臭氧分解

P

pach- 【构词成分】厚（度）
pachometer 测厚仪
pack 包装，包裹；包装，压紧，捆扎，挑选，塞满
pack annealing 装箱退火，叠式［堆垛，成叠］退火
pack carburizing 固体渗碳
pack hardening 装箱渗碳硬化
pack heating furnace 叠板加热炉
package 包，包裹；一揽子的；打包，将……包装
packaged 包装过的，小型的，袖珍的，典型的，综合的，成套的
packed bed 填充床
packed column 填充柱，填充塔
packing 填充物，包装
packing agent 渗碳剂
packing density 堆积密度
packing factor 堆积因子
packing list 装箱单，包装单
packingless 不能密封的，无密封的
packless 无填料［填充，密封，衬垫］的，未填实的，疏松的
pad 衬垫，护具；填补［补，塞］
pad printing 移印
padding 揩涂，填料，垫料，芯，浸染，使平直，使均匀
paddle 搅棒，浆（状物）
paint 涂料，油漆
paint abraded 喷油［漆］，擦花
paint circulating system 涂料循环输漆系统
paint filler 油漆灌装机，油漆填料［底层］
paint film appearance 涂膜外观
paint film softness 漆膜柔软
paint fog 漆雾
paint insufficient coverage 喷油反底，露底色
paint mixing 调漆
paint process 涂装工艺，油漆工艺
paint remover 脱漆剂，洗漆水
paint sprayer 喷漆器，喷漆枪
paint system 涂装体系
paint thinner 涂料用稀料，涂料稀释剂
paintbox 颜料盒，绘画箱
paintbrush 漆刷
paintcoat 涂层，漆层
painted 着色的，刷上油漆的，彩色的
painting 涂漆，着色，颜料，油漆
painting make-up 补漆
painting peel off 脏污，掉漆，甩油
painting robot 涂装机械手
painting sepctification 涂装技术要求
painting smear 喷油拖花
paint stripper 油漆去除剂，除漆剂
pair ion 离子对
paired ion chromatography (PIC) 离子对色谱法
pale 苍白的，暗淡的，浅色的，微弱的；变淡
pale blue 淡蓝色
pale (o) - 【构词成分】古，原始，旧
palette 调色板，颜料
palid 巴里特合金，铅基轴承合金
palingenesis 再生（作用），新生
palirrhea 回［反］流，再度漏液
palium 铝基轴承合金
palladic 钯的，含钯的（尤指含4价钯的）
palladium 钯
palladium black 钯黑
palladium chloride 氯化钯，二氯化钯
pallador 铂钯热电偶
pallet 托板，调色板
palliative 减尘剂，防腐剂；缓和的，减尘的

palm butter 棕榈油
palmitate 棕榈酸盐，棕榈酸酯，棕榈酸，十六酸
palmitic acid 棕榈酸，十六酸，软脂酸
palmitoyl- 【构词成分】棕榈酰（基），软脂酰（基）
pan mill 碾磨盘，磨石，碾碎机
panel board 镶块，配电板，仪表板
panel 仪表板，嵌板
panlite 聚碳酸酯树脂
pano- 【构词成分】全，总，泛
panoramic 全景的
panoramic exposure 全景曝光
panradiometer 全波段辐射计
pantograph 放大尺，缩放仪，缩图仪，比例画图仪器
pantomorphism 全对称性，（结晶）全对称现象，全形性
pantoscopic 视域广阔的，广角的
pantothenate 泛酸盐
pantothenic 泛酸的
paper chromatography 纸色谱法，纸层析法，纸色谱
paper tack stick 黏性纸胶带
par (a)- 【构词成分】侧，并，外，旁，顺，超，类，似，拟，异常，对位，聚（合），仲，副
para position 对位
para-compound 对位化合物
parabolical 抛物线的
paracresol 对甲酚
paracril 丁腈橡胶
paracrystal 次［仲］晶，不完全结晶
paracrystalline 次晶的，亚晶状的，类结晶的
paraffin 石蜡，链烷烃，硬石蜡；用石蜡处理，涂石蜡于……
paraffin series 石蜡系，石蜡族烃
paraffin wax 石蜡，固体石蜡
paraffinaceous 石蜡的
paraffinic 石蜡的，链烷的

paraffinicity 链烷烃［石蜡］含量
paragutta 合成树胶，巴西橡胶，假橡胶
parallel 平行，对比；平行的，类似的，相同的；使……与……平行
parallel and level 平齐
parallel scan 平行扫查
paralleled 并行的，平行的，并联的
parallelehedra 平行面体
parallelepiped 平行六面体
parallelism 平行度，平行，类似，对应
paramagnetic material 顺磁性材料，顺磁物质
paramagnetism 顺磁性，常磁性
parameter 参数，系数，参量
parameter extraction 参数萃取，参数提取
paramorph （晶体）同质异形，副象
paramorphic （晶体）同质异形的
paramorphism 同质异晶，同质假象，同质异形性，全变质作用
parasitic echo 干扰回波
paratactic 并列的
paratellurite 亚碲酸盐
paratungstate 仲钨酸盐
paraxylene 对二甲苯
parch 烤，烘；使干透，焦干，烤干
parch crack 烤裂，拉裂，干裂
pare 消减，削皮，剪，修掉（边等）
parent metal 母材，基层金属
pareto diagram 柏拉图，排列图
parkering 磷酸盐处理，磷化处理
parkerise 磷酸盐被膜［保护膜］（防锈）处理
parol 石蜡燃料
part of billion (ppb) 十亿分之几，微克/升
part of million (ppm) 百万分之几，毫克/升
partial 局部的，部分的
partial condensation 部分冷凝
partial load 部分负荷，分载，局部负载
partial potential 偏势

partial pressures 分压力
partial reaction 部分反应
partially 部分地，局部地，不完全地
partially elastic 部分弹性
partible 可分的
particle 质点，粒子，微粒
particle diameter 颗粒（粒子）直径，粒径
particle electrophoresis 粒子电泳
particle reinforced composition 颗粒增强复合材料
particle size 粒度，颗粒大小
particle size analysis 粒径分析，粒度分析
particle size analyzer 粒度分析仪，粒度分析器
particulate 微粒，微粒状物质；微粒的
particulate copper 颗粒性铜，散式铜
particulate solids 粉碎的固体颗粒，（催化剂的）粉碎固体粒子
parting 夹层，分支，分离［脱模］剂；分开的
parting agent 模型润滑剂，分型［脱模］剂
partition 划分，分开；分割，分隔，区分
partition chromatography 分配色层法，分溶层析法
partition coefficient 分配系数
passage 通路
passivant 钝化剂
passivate 使钝化
passivation 钝化（作用），形成保护膜
passivation effect 钝化效应
passivation potential 钝化电位
passivator 钝化剂，减活剂
passive 被动的，消极的，钝态的，不活泼的
passive state 钝态
passive-active cell 钝化-活化电池
passivity 钝态，钝化
paste 膏，胶，糊状物；涂胶
pasted 膏的（胶的，糨糊的）

patching （炉衬）修补［理］
patching material 修补材料
patchwork 混杂物，修补工作，拼凑的东西
patchy 不调和的，不规则的，杂凑的，质地不均匀的
patent 专利；专利的，新奇的，显然的；（钢丝）韧化处理，铅淬火
patenting 铅浴处理，铅淬火，线材的拉丝后的退火处理
patentizing 钢丝韧化处理，铅淬火
patina 铜绿［锈］，（金属或矿物的）氧化表层，任何外面之物
patinated 古色古香的，生了锈的，布满铜绿的
patination 生锈，布满铜绿
patinous 生绿锈的，有锈的
patronite 绿硫钒矿，绿硫钒石
patten 木质磨块，挡板
pattern 模式，图案，样品；模仿，以图案装饰
pattern cracking 网状裂纹
pattern draw 脱［起］模
pauci- 【构词成分】（微）少
paucidisperse 少量分散
paucity 缺乏，少数，少量（许），贫乏
peak 峰值；最高的，最大值的
peak current 峰电流
peak energy 谱峰能量，峰值能量
peak fitting 峰拟合
peak inverse voltage 最大反向电压，反峰电压
peak load 峰值负荷，最大负载
peak potential 峰电位
peak width at half height 半峰宽
peak width 峰宽
peaky 多峰的，尖峰的
pearl 珍珠，珍珠色；镶珍珠的，珍珠状的；使成珠状，珍珠装饰，使呈珍珠色
pearl chrome 珍珠铬
pearl nickel 珍珠镍

pearl glue 珍珠胶
pearl white 锌钡白
pearled 用珍珠装饰的，珍珠似的，有珍珠色彩的
pearlite 珠光体，珠光体组织
pearlitic 珠光体的
pectinose 果胶糖，阿拉伯糖
pectization 凝结，胶凝，胶化
pedi- 【构词成分】脚
pedial 单（晶）面的
pedion 单面晶
peel 皮；剥，剥落，削
peel off 剥离［去］，脱掉
peel strength 剥离强度
peeling 剥落，起皮
peeling test 剥离试验
peen 锤头；用锤头敲打
peening 锤击，轻敲，锤平，喷珠（丸）硬化
pelletize 把……制成小球，制粒，做成丸状
pencil hardness test 铅硬度试验
pendent 下垂的，悬垂的
pendent drop apparatus 悬滴法表面张力测定仪
pendular 摆动的，钟摆运动的
pendulate 摆动，摆振，振动
pendulous 下垂的，悬垂的，摇摆的
penetrability 穿透性，透入性，透过率，穿透能力，渗透性
penetrable 可穿透的，可渗透的，能透过的，不密封的
penetrameter 透度计
penetrant 渗透剂，渗透液；渗透的
penetrant flaw detection 渗透探伤
penetrate inspection 渗透探伤法
penetrate 渗透，透［穿，渗，贯，陷，进］入
penetrating 渗透的
penetrating agent 渗透剂

penetration 侵［贯，透，穿］入，渗［穿］透
penetrative 渗透的，有穿透力的
penetrometer 针穿硬度计，射线透度计，透光计，稠密度计
pennate 有翼的，羽状的
penniform 羽状的
pent (a) - 【构词成分】五，戊
pentane 戊烷
pentanedioic acid 戊二酸
pentanediol 戊二醇
pentanoate 戊酸［盐，酯，根］
pentanoic acid 戊酸
pentanol 戊醇
pentavalence 五价
pentavalent 五价的
pentylene 戊二烯
peptizate 胶溶体
peptization 胶溶作用，分解作用，解胶
peptizator 胶化剂，胶溶剂
peptizing agent 胶溶剂
peptize 使胶溶，塑解
peptizer 塑解剂，胶溶剂
pentoxide 五氧化物
per- 【构词成分】通（过），遍及，完全，极，超，甚，过，高
peracid 过酸，高酸
peracidity 过酸性
peralkaline 过碱性
perbasic 高碱性的
perbenzoic acid 过苯（甲）酸
percentage 百分比，百分率，百分数
percentage of elongation 延伸率
percentage reduction of area 断面收缩率
perchlorate 高［过］氯酸盐（或酯）
perchloroethylene 四氯乙烯，全氯乙烯
perchloric acid 高氯酸，过氯酸
perchloride 高氯化物，过氯化物
perchloromethane 四氯化碳
percolate 滤过液，渗出液；使渗出［过滤］，过滤，渗出，浸透

percolating filter 渗滤池，渗透滤器
percolation 过滤，浸透[滤，漏]
percolation apparatus 渗滤器[仪]
percolation ratio 渗透率，渗漏比
percolator 渗透器，渗滤浸出器，过滤器，滤池
percrystallization 透析结晶（作用）
percussive 冲[撞]击的
percussive boring 冲击钻孔[探]
perdistillation 透析蒸馏作用
perdurability 持[耐]久性，（延续）时间
perdurable 永[持，耐]久的
perduren 硫化橡胶
perester 过酸酯
perfect combustion 完全燃烧
perfect crystal 完整晶体，完美晶体
perfect mixing 完全混合
perfectness 完美，分毫不差，精通
perflation 通风，换气
perflectometer 反射显微镜，反射头
perfluorination 全氟化作用
perfluoroalkylation 全氟烷基化
perforate 穿孔于，打孔穿透，在……上打齿孔
performance 使用性能，操作效能，特征（曲线）
performance data 性能[运行]数据，工作特性，动态参数
performance depreciation 性能下降
performance figure 性能指标，性能数字，质量指数效率
performance parameter 性能参数
perfume 香水，香味；洒香水于……，使……带香味，散发香气
perfuse 灌注，洒遍，散布
perfusive 撒遍的，易散发的，能渗透的
pergameneous 质地（或纹理等）似羊皮纸的，羊皮纸的
pergamyn 羊皮纸
pergamyn paper 耐油纸

perhalogenation 全卤化（作用）
perhydrate 过水合物
perhydro- 【构词成分】氢化
perhydrol 强双氧水（含30％过氧化氢）
period 周期
periodate 高碘酸盐
periodic 周期的，定期的，循环的，高碘的
periodic acid 高碘酸
periodic law 周期律
periodic precipitation 间歇沉淀
periodic purification 定期净化
periodic reaction 周期反应
periodic reverse current 周期换向电流，周期反向电流
periodic reverse plating 周期换向镀
periodic reverse pulse plating 周期换向脉冲电镀
periodic table 周期表
periodicity 周期性
peripheral 外围设备，辅助设备；外围的，次要的，边缘的，外表面的
peripheral velocity 圆周速度，周缘速度，轮缘速度
peripheric 外围设备；外围的，周围的
perishable 容易腐坏的东西；易腐坏的，易毁灭的，会枯萎的
peristasis 蠕动
peristaltic 蠕动的
peritectic 包晶；包晶的，转熔的
peritectic phase diagram 包晶相图
peritectic reaction 包晶反应
perlit 珠光体铸铁
perlite 珠光体，珍珠岩
perlitic structure 珍珠构造，珍珠岩结构，珠光结构
permag 清洁金属用粉
permanence 持久性
permanent 永久的，永恒的，不变的
permanent magnet 永久磁铁
permanent magnetic material 硬磁材料，

永磁材料
permanent red 永久红
permanent set 永久变形，残余变形
permanent stability 长期稳定性，耐久性
permanently 永久地，长期不变地
permanganate 高锰酸盐
permanganate demand 高锰酸盐需氧量
permanganate method 高锰酸盐［钾］（滴定）法
permanganate titration 高锰酸盐滴定法
permanganic 含最高价锰的
permanganic acid 高锰酸
permanganometry 高锰酸盐滴定法
permeability 渗透性，透磁率，磁导率，弥漫
permeable 能透过的，有渗透性的，不密封的
permeable membrane 可透膜
permeameter 渗透计，磁导仪，渗透性试验仪
permeance 磁导，浸透，透过
permeate 渗透，透过，弥漫
permeation 渗［浸］透，贯穿，透过
permeation coefficient 渗透系数
permittance 电容性电纳，电容，静电电容
permittivity 介电常数，电容率
permeameter 透气性试验仪
permolybdate 过钼酸盐
permselective 选择性渗透的
permselective membrane 选择性渗透膜
permselectivity 选择渗透性
permutability 可置换性，转置性，换排性
permutable 可排列的，能交换的
permutation 置换，变更，取代
Permutit （天然或人造）沸石，一类离子交换树脂
permutite 软水砂，人造沸石，滤砂
permutoid 交换体
permutoid reation 交换体反应
pernicious 有害的，恶性的，致命的，险恶的
peroikic 多主晶的
peroxide 过氧化物；以过氧化物漂白的；以过氧化物处理
peroxide effect 过氧化物效应
peroxide treatment 过氧化物处理
peroxides and superoxides 过氧化物和超氧化物
peroxidize 过氧化，使变为过氧化物
peroxyl 过氧化氢
peroxysulfate 过（氧）硫酸盐
perpendicular 垂直的，正交的，垂线
perspective 远景，透视图，透镜；透视的
perspiration 出汗，分泌，蒸发，排出
persulfate 过硫酸盐
persulfuric acid 过硫酸
persulphuric acid 过硫酸
pertaining 附属的，与……有关的
pervade 遍及［布］，弥漫
pervaporation 渗透蒸发，全蒸发
perveance 导流系数，电子管屏导数
pervial 透水的，通透过的
pervious 能被通过的，能接受的，可渗透的
perviousness 透水性，透过性，渗透性
petrol 汽油，挥发油
petrolatum 矿脂，石蜡油，凡士林油
petrolatum album 白凡士林，白矿脂
petroleum 石油
petroleum benzine 石油醚
petroleum distillation 石油蒸馏
petroleum ether 石油醚，石油精
petroleum jelly 凡士林，矿油
petroleum spirit 汽油，石油精
petrolic 石油的，汽油的，从石油中提炼的
petroliferous 产油的，含石油的
peucine 沥青，树脂
peucinous 树脂的
pexitropy 冷却结晶作用
pH indicator pH指示剂

pH meter   pH 计
pH standard solution   pH 标准溶液
pH test paper   pH 试验纸
pH value   pH 值
phacoid   透镜状的，透镜体
phase   相
phase boundary   相界
phase boundary potential   相界电位
phase diagram   相图
phase equilibrium   相平衡
phase rule   相律
phase separation   相分离
phase transfer catalysis   相转移催化作用
phase transformation   相变
phase transformation toughening   相变增韧
phenate   石炭酸盐，苯酚盐
phenethyl   苯乙基
phenethylene   苯乙烯
phenixin   四氯化碳
phenol   石炭酸，苯酚
phenol resin   酚醛树脂
phenolic   酚醛树脂；酚的，酚醛树脂的，石炭酸的
phenolic coating   酚醛树脂涂料
phenolic paint   酚醛漆
phenolic plastics   酚醛塑料
phenolic resin   酚醛树脂
phenolic resin paint   酚醛树脂涂料
phenolic resin varnish   酚醛清漆
phenolphthalein   酚酞
phenoplast   酚醛塑料
phenolsulfonate   苯酚磺酸盐
phenolsulphonate   苯酚磺酸盐
phenotype   表型，表现型，显型，混合型
phenyl   苯基
phenylalanine   苯基丙氨酸
phenylamine   苯胺
phenylbenzene   联苯
phenylene   亚苯基
phenylethane   苯乙烷，乙苯
phonon   声子

phoresis   电泳现象
phos-   【构词成分】光
phosgenation   光气化，光气化作用
phosgene   光气，碳酰氯
phosph-   【构词成分】磷
phosphate   磷酸盐
phosphate coating   磷化膜，磷酸盐处理法
phosphate process   磷酸盐法
phosphatic   磷酸盐的，含磷酸盐的
phosphatic deposit   磷质沉积
phosphatidate   磷脂酸（盐，酯，根）
phosphatide   磷脂
phosphatidylcholine   卵磷脂，磷脂酰胆碱
phosphating   磷化，（金属表面）磷酸盐（防锈）处理
phosphating chemicals   磷化处理剂
phosphatization   磷化，磷酸盐化
phosphatizing   磷化处理，磷化，渗磷
phosphide   磷化物，磷脂
phosphinate   亚膦酸盐
phosphine   磷化氢，三氢化磷
phosphinyl   氧膦基，磷酰基
phosphite   亚磷酸盐（酯）
phospho-   【构词成分】磷酰，磷酸（基）
phosphoamide   磷酰胺
phosphoglyceride   磷酸甘油酯
phospholipid   磷脂
phosphomolybdate   磷钼酸盐
phosphonation   膦酸化（作用）
phosphor bronze   磷青铜
phosphor   荧光物质，磷光体，磷光剂
phosphor removal   去磷，脱磷
phosphor-copper   磷铜
phosphorate   使发出磷光，加磷，使和磷化合
phosphoresce   发出磷光
phosphorescence   磷光（性），荧光（现象）
phosphorescent   发出磷光的，磷光性的
phosphorescent materials   磷光材料
phosphoreted   与磷化合的，含磷的
phosphoric   磷的，含（五价）磷的

phosphoric acid 磷酸
phosphoric acid anhydride 磷酐
phosphorimetry 磷光分析，磷光光度法
phosphorization 磷化作用，增磷
phosphorize 加磷，使发出磷光
phosphorized rolled copper 磷铜球
phosphoro- 【构词成分】磷（的）
phosphorous （亚）磷的，含（三价）磷的
phosphorous acid 亚磷酸
phosphorous anhydride 亚磷酸酐
phosphorous pentoxide 五氧化二磷
phosphorus 磷
phosphorylation 磷酸化（作用）
phot- 【构词成分】光（致，敏），摄影［像］，光电，光子，光化（学的）
photelometer 光电比色计
photetch 光蚀，光刻技术
photic 光的，透光［感光，发光］的
photo- 【构词成分】光，光电，照相
photoabsorption 光吸收，光致吸附，光电吸收
photoactinic 能产生光化作用的，发光化射线的
photoactivate 使光敏化，用光催化
photoactive 光敏的，感光性的，光活化的
photoactive substance 感光物质，光敏物质，光活物质
photoacoustic 光声的
photoacoustic spectroscopy 光声光谱学
photoadsorption 光吸附，光致吸附
photoaging 光老化
photoanalysis 光分析，光电分析
photoassociation 光缔合
photoautotrophic 能光合自养的
photoautotrophic bacteria 光能自养菌
photocatalysis 光催化（作用），光化学催化，光接触作用
photocatalyst 光催化剂
photocathode 光电阴极，光电发射体
photochemical 光催化学物；光化学的
photochemical catalysis 光催化

photochemical cell 光化学电池
photochemical cleaning equipment 光化学洗涤［清洗］设备
photochemical equilibrium 光化学平衡
photochemical reaction 光化反应
Photochemical rearrangement 光化学重排
photochemistry 光化学
photocolorimeter 光比色计，光斑比色计
photocolorimetry 光比色法，光斑比色法
photoconduction 光电导（率，性）
photoconductive 光电导的，光敏的
photoconductive cell 光传导电池，光电导管，光敏电阻
photoconductivity 光电导率
photocreep 光蠕变
photocrosslinking 光致交联
photocurrent 光电流
photocurrent carrier 光电载流子
photodecomposition 光解作用
photodegradable 可光降解的
photodegradation 光降解，光降解作用
photodepolarization 光去极化
photodestruction 光裂解，光化裂解
photodiffusion 光扩散，光致扩散
photodimer 光二聚物
photodimerization 光二聚作用
photodiode 光电二极管
photodiode array detector 光电二极管检测器
photodissociation 光致离解［分离，分解］，光解，光化学离解
photoeffect 光（电）效应
photoelectric 光电的
photoelectric cell 光电池［管］
photoelectric color comparator 比色仪，光电比色仪
photoelectric colorimeter 光电比色计
photoelectric colorimetry 光电比色法，光电比色分析法
photoelectric conversion material 光电转换材料

photoelectric effect 光电效应
photoelectric spectrophotometer 光电分光光度计，光电分光计
photoelectric tube 光电管
photoelectrochemistry 光电化学
photoelectron 光电子
photoelectroluminescence 光电发光，光控电场（致）发光
photoelectrolytic 光电解的
photoemission 光电发射，光（致）发射
photoemissive 光（电）发射的
photoemissivity 光发射能力
photoengraving 光刻蚀，照相制版，光刻
photoesthetic 感光的，光觉的
photoexcitation 光致激发
photoextinction 消光，比浊分析法
photohalogenation 光卤化（作用）
photoinduction 光诱导，光感应
photoisomerization 光异构化
photolabile 对光不稳的，光致不稳定的，不耐光的
photolithography 光刻［蚀］法
photoluminescence 光致发光，光激发光
photoluminescence method 光激发光法
photolysis 光分解，光解，光解作用
photolyte 光解质（物）
photolytic 光解的
photomagnetic 光磁的
photomagnetic effect 光磁效应
photomagnetism 光磁性
photomask 光罩，光掩模
photometer 光度计
photometric 测光的，光度计的，光度测定的
photometric titration 光度滴定
photomicrograph 微观照片
photon 光子，辐射量子，见光度
photonegative 负趋光性的，负光电导的
photooxidant 光氧化剂
photooxidation 光（致）氧化（作用），感光氧化作用

photoparametric 光参数的
photoperiodicty 光周期（性，现象）
photophoresis 光泳现象，光致漂移，光致迁动
photophosphorylation 光（合）磷酸化（作用）
photopolymerisable 光聚合的
photopolymerization 光聚合，光致聚合作用
photoprocess 光学处理，光学加工
photoproduction 光致产生，光致作用
photosensitive 光敏的，感光性的
photosensitiveness 感光度，感光性，光敏性
photosensitization 光敏化
photosensitize 使感光［光敏］
photothermal 辐射热的，光热的
photothermal spectroscopy 光热光谱
photothermy 光［辐射］热作用
photovoltaic 光电伏打的，光电的
phragmoplast 成膜体
phthalic 邻苯二甲酸的，酞酸的
phthalic acid 苯二甲酸
phthalic anhydride 邻苯二甲酸酐
phthalocyanine 酞化青染料，苯二甲蓝染料，酞菁染料
physical adsorption 物理吸附
physical analysis 物理分析
physical and mechanical properties 物理及机械性能
physical chemistry 物理化学
physical property analysis 物性分析
physical testing 物理测试
physical vapor deposition (PVD) 物理气相沉积法
physicochemical analysis 物理化学分析，理化分析
physiological chemistry 生理化学
physiological salt solution 生理盐溶液
physisorption 物理吸附

phytic acid 植酸，肌醇六磷酸
picked 精选［挑选］的，尖的，有尖峰的
pickle （浸渍用）盐水，（清洗金属表面用）酸洗液，稀酸液；酸洗［浸，蚀］，浸泡
pickler 酸洗设备［装置］，酸洗液
pickling 强侵蚀
pickling bath 酸洗池
pickling solution 酸洗液
pickling tub 酸洗池
picotite 铬尖晶石
picrate 苦味酸盐
picric 苦味酸的
picric acid 苦味酸
pictogram 象形图，图表［解］
pictorial 绘画的，形象化的，有插图的
pied 杂色的，斑驳的
pierce 冲［穿，锥，钻］孔
piercing 冲孔加工；刺穿的，锐利的
piercing die 冲孔模
pierce punch 冲孔冲头
piesimeter 压力计
piezo- 【构词成分】压（力，电）
piezocaloric 压热（的）
piezocrystallization 压结晶作用，加压结晶，压致结晶
piezodialysis 压力透析，加压渗析
piezodielectric 压电介质的
piezoelectric 压电的
piezoelectric effect 压电效应
piezoelectric material 压电材料
piezoelectric stress constant 压电应力常数
piezoelectric voltage constant 压电电压常数
piezometric 量压的，压力计的
piezoresistance 压电电阻
piezoresistivity 压电电阻率，压电电阻效应
pigment 颜料
pigment content 颜料分［含量］

pigment sedimentation 颜料沉降
pigment volume concentration 颜料体积比率，颜料体积浓度
pigmental 颜料的
pigmentation 染色，着色，颜料沉积（作用）
pile 堆，大量；累积
pillow 垫［枕，轴承］座，垫板［块］
pilot 试点的
pilot plant test 中试
pimpling 凸起，粗糙度
pin 销钉
pin point 针点型，（多孔镀铬的）点状孔隙
pinacoid 轴面，平行双面式
pinacol 频哪醇
pinch 挤压（变形），收缩效应，夹痛；捏［夹，捻，挤］，夹紧，紧［挤］压
pinchbeck 金色黄铜，廉价仿制品；冒牌的，赝制的，金色铜制的，波纹管状的
pinched 压紧的，收缩的
pincher 钳子
pinhead 针头，微不足道的东西
pinhead blister 微［针，气］孔
pinhole 针［销，栓，塞，冲，凿，穿，小］孔
pinhole test 小孔试验
pink 粉红色；粉红的
pinkish 略带桃色［粉红色］的，浅桃色
pint 品脱
pipe 移液管
pipelining 流水线，管道安装；用管道输送
piperine 胡椒碱
piperonal 3,4-二氧亚甲基苯甲醛，洋茉莉醛，胡椒醛
pipet 吸移管，吸移管；用移液管移
pipet support 吸移管架
pipette 移液管，吸移管；用移液器吸取
piping 管道敷设，管道系统
pipy 管形的，管状的

piston 活塞
pit 坑点，麻点；凹陷，起凹点
pitch （纹槽）间距，节［螺，齿，辊，纹，铆，栅］距，斜度
pitch gauge 螺距规，节距规，螺纹样板
pitchblack 深黑色，漆黑的
pitchy 漆黑的，沥青的，黏性的
pitted 把……放进坑内；有麻点的，有凹痕的
pitting （金属）点蚀，麻点状［局部］腐蚀，剥蚀，点状疏松
pitting breakdown potential 点蚀击穿电位
pitting corrosion 点状腐蚀，孔腐蚀
pitting factor 点蚀系数
pitting potential 点蚀电位，孔蚀电位
pitting test 点蚀试验
pivot 轴，中心点，旋转运动；枢轴的，关键的；以……为中心旋转，把……置于枢轴上，在枢轴上转动，随……转移
plain 平的，简单的，清晰的；清楚地，平易地
plain carbon steel 普通碳钢，碳素钢
plain die 简易模，简单模
plain sedimentation 自然沉淀
planar 平面的，二维的，平坦的
planar magnetron sputtering system 平面磁控管溅镀系统
plane 平面；平的，平面的；刨平，用刨子刨
plane chromatography 平板色谱法
plane defect 面缺陷
plane electrode 平板电极
plane strain fracture toughness 平面应变断裂韧性
plane strain 倒角应力，平面应变（变形）
planeness 平面度，平整度
plani- 【构词成分】平（面）
planiform 平面的，扁平形的
planing 平刨，正交法；刨的，正交法的
planing machines 刨床，刨削机
planish 压平，打平，（金属等的）精轧

plano- 【构词成分】平（面），流动
plasma 等离子体
plasma assisted controlled thinning method 等离子辅助控制薄膜化加工法
plasma cleaning 等离子体清洗
plasma cleaning equipment 等离子体洗涤设备
plasma damage 等离子体损伤
plasma doping system 等离子体掺杂系统
plasma enhanced CVD system 等离子体增强化学气相沉积系统
plasma electrolytic oxidation 等离子电解氧化
plasma heat treatment 等离子热处理
plasma nitriding 离子氮化，离子渗氮
plasma oxidation furnace 等离子体氧化系统
plasma panel 等离子体显示板（或屏面）
plastic 塑料制品，整形，可塑体；塑料的，可塑的
plastic deformation 塑料变形，可塑成型
plastic making 塑性成形
plastic wash bottle 洗瓶，塑料洗瓶
plasticity 塑性，可塑性［度］，适应［柔顺，柔软］性，黏性
plasticity coefficient 塑性系数
plastics 塑料
plastics container 塑料蓄电池槽
plasticization 塑化，增塑作用
plasticizer 增塑剂
plasticine 橡皮泥，塑像用黏土
platability 可镀性
plate mark 模板印痕
plate bending rolls 卷板机，弯曲滚板机
plated 镀……的，电镀的
plated metal 电镀金属，沉积金属
platelike 层状的，板状的
platform 平台；平台式的
platform balance 托盘天平，台秤
platina 铂，白金
platine 装饰用锌铜合金

plating 电镀，镀层
plating on aluminum 铝材电镀
plating on plastics 塑料电镀
plating rack 挂具（夹具）
plating tank 镀槽
platini- 【构词成分】铂，白金
platinic 白金的，含白金的
platinic acid 铂酸
platinic chloride 氯化铂，四氯化铂
platiniferous 含铂的，产铂的
platinoiridita 铂铱
platinous 亚铂的
platinum 铂，白金
platinum black 铂黑
platinum metals 铂类金属
platy 板状的，扁平状的
platy- 【构词成分】扁平，宽，阔
plei (o) - 【构词成分】多
pleiotropy 多效性，基因多效性，多向性
plenum 充满，增压
plenum box 充气箱
plenum ventilation 压力通风，压入式通风
pleochroic 多色的
pleochroism 多向色性，多色现象
pleomorphic 多形的，多晶的
pleomorphism 多形性，多型现象，同质多晶
pleomorphous 多形的
plesiomorphic 形态相似的
plesiomorphism 形态相似
plesiomorphous 形态相似的
pliability 柔软，易曲折，可挠［弯，锻，塑］性
pliability test 弯曲性试验
pliable 柔韧的，柔软的，圆滑的，易曲折的
plication （细）褶皱，皱纹
plier 镊钳，钳，镊子
plio- 【构词成分】多，更多
plotting 测绘，标图
ployester 聚酯纤维，涤纶

plucking 剥痕
plug 插头，塞子，栓；插入，塞住，接插头
plumb 垂直，铅锤；垂直的；使垂直；正，垂直地
plumbagine 石墨（粉）
plumbaginous 石墨的，似石墨的，含石墨的
plumbate 铅酸盐；高铅酸盐四价铅的
plumbean （正）铅的
plumbeous （正，似，含）铅的，铅灰色的
plumbic 铅的，含四价铅的
plumbiferous 含铅的，产铅的
plumbism 铅中毒
plumbite 铅酸盐，亚铅酸盐
plumbous （亚，二价）铅的
plumose 羽毛状的，有羽毛的
pluvious 多雨的
pneumatic 气胎；气动的，充气的，有气胎的
pneumatic power tools 气动工具
pneumatic pressure 气压
pneumatic test 气压试验
pneumatolysis 气化
pock 麻点
pock mark 痘斑，麻点
pocket 小型的，袖珍的
pocket dosimeter 袖珍剂量计，携带式放射线剂量计
pocket type plate 袋式极板
poecilitic 嵌晶结构的
poecilosmoticity 变渗［透压］性
poikilitic 嵌晶结构的
point defect 点缺陷
point inspection 点检
pointing needle 方向针
pointy 尖的，非常尖的
poison 毒药，毒物，抑制剂；有毒的；污染，使中毒，放毒于
poisonous 有毒的，恶毒的

Poisson ratio 泊松比，横向变形系数
polar 极面，极线；极地的，两极的，正好相反的
polar bond 极性键
polar group 极性基团
polar molecule 极性分子
polari- 【构词成分】极
polarity molecules 极性分子
polarizability 极化度，极化率，可极化性
polarizable 极化的
polarizable electrode 极化电极
polarization 极化，极化作用
polarization curve 极化曲线
polarization potential 极化电位
polarization resistance 极化电阻
polarized 极化的，偏振的
polarogram 极谱图
polarograph 极谱仪
polarographic 极谱法的
polarographic cell 极谱池
polarographic current 极谱电流
polarographic maxima 极谱极大
polarographic titration 极谱滴定法，安培滴定法
polarographic wave 极谱波
polarography 极谱法
pole 极，极柱
polish 磨光，擦亮，上光剂，擦亮剂；磨光，使发亮
polisher 磨光器
polishing 抛光
polishing anode dyeing 磨光阳极染色
polishing compound 抛光剂
polishing lathe 抛〔磨，擦〕光机
polishing mark 抛光痕
polishing paste 抛光膏，磨光〔擦光〕剂
pollute 污染，玷污
polluted 受污染的，被玷污的
pollution 玷污，污染
poly- 【构词成分】多，聚，复，重

polyacid 缩多酸，多元酸；多酸的
polyacrylamide 聚丙烯酰胺
polyacrylate 聚丙烯酸酯
polyacrylic acid 聚丙烯酸
polyaddition 逐步加成聚合作用
polyamide 聚酰胺
polyamide resin 聚酰胺树脂
polyamine 聚（酰）胺，多胺
polyatomic molecules 多原子分子
polybasic acid 多元酸
Polybrominated biphenyls 多溴联苯
polybrominated diphenyl ethers 多溴联苯醚
polybutadiene 聚丁二烯
polycarbonate 聚碳酸酯
polychromate 多色物质
polychromatic 多色的，色彩变化的
polychrome 多色画，彩色物品；多彩的，多色装饰的；用多色装饰
polychromic 多彩艺术品；彩饰的
polychromy 色彩装饰，多色性
polycondensate 缩聚物
polycondensation 缩聚作用
polycrystalline 多晶的
polycrystallinity 多晶性
polycyclic 多环的，多相的
polydisperse 多分散的，杂散的
polydispersity 多分散性，聚合度分布性，杂散性
polyhedron 多面体
polyelectrolyte 聚合电解质
polyelectrolyte effect 聚电解质效应
polyepoxide 聚环氧化物
polyenic 多烯的
polyester 聚酯
polyester coatings 聚酯漆喷涂
polyester putty 聚酯腻子，聚酯油灰
polyester resin 聚酯树脂
polyesteramide 聚酰胺酯
polyesterification 聚酯（化，作用）
polyether 聚醚，多醚

polyethylene 聚乙烯
polyethylene glycol 聚乙二醇，聚氧乙烯
polyethylene oxide 聚氧化乙烯，聚环氧乙烷
polyflon 聚四氟乙烯树脂
polyfluorated 多氟化的
polyfuctional 多官能团的，多机能的，多作用的，多重（性）的
polyglycine 聚甘氨酸，多聚甘氨酸
polyglycol 聚乙二醇
polygonal 多边形的，多角形的
polygonization 多边化，多边形化
polyhydrate 多水合物
polyhydric 多羟的，多羟基的
polyimide 聚酰亚胺
polylaminate 多层的
polymer 聚合物
polymer homologue 同系聚合物
polymer surfactant 高分子表面活性剂
polymeric micelle 聚合物胶束
polymerization 聚合
polymerization accelerator 聚合促进剂，聚合加速剂
polymerization degree 聚合度
polymerize 使……聚合，聚合
polymerized 聚合的
polymerous 聚合状的
polymethacrylate 聚甲基丙烯酸酯
polymolecular 多分子的
polymorphic 多态的，多形的，多形态的，多晶形的
polymorphism 多晶型性
polymorphous 多形的，多形态的，多功能的
polynuclear aromatic hydrocarbons 多环芳烃
polyol 多元醇，多羟基化合物
polyoxy 聚氧
polyoxyethylene octylphenol ether 辛基酚聚氧乙烯醚
polyoxymethylene 聚甲醛

polyphenol 多元酚
polypropylene 聚丙烯
polypropylene oxide 聚环氧丙烷
polysaccharide 多糖，多聚糖
polysorbate 聚山梨醇酯，多山梨酸酯，聚氧乙烯脱水山梨醇酯
polystyrene 聚苯乙烯
polystyrene resin 聚苯乙烯树脂
polysulfone 聚砜
polytetrafluoroethylene 聚四氟乙烯
polythene 聚乙烯
polyurethane 聚氨酯
polyurethane coatings 聚氨酯漆膜
polyurethane paint 聚氨酯漆
polyurethane varnish 聚氨酯清漆
polyvalent 多价的
polyvinyl 聚乙烯化合物；乙烯聚合物的
polyvinyl acetate 聚醋酸乙烯
polyvinyl alcohol 聚乙烯醇
polyvinyl chloride 聚氯乙烯
polyvinyl chloride resin 聚氯乙烯树脂
polyvinyl ether 聚乙烯醚
polyvinylamine 聚乙烯胺
polyvinylbutyral 聚乙烯醇缩丁醛
polyvinylchloride 聚氯乙烯
polyvinylfluoride 聚氟乙烯
polyvinylidenechloride 聚偏二氯乙烯
polyvinylpyrrolidone（PVP） 聚乙烯吡咯烷酮
pore 针孔
porosity 孔隙，孔隙率，有孔性，多孔性
porous 多孔渗水的，能渗透的，有气孔的
porous anodic oxide film 多孔阳极氧化膜
porous glass 多孔玻璃膜
porous layer 多孔层
portable 手提的，便携式的，轻便的
positional 位置的，地位的
positional isomers 位置异构体
positive 正的
positive electrode 正极
positive feedback 正反馈

positive ion 阳离子
positive plate 阳极板
positive valency 正化合价
post emulsification 后乳化
post treatment process 后处理过程
postplating 镀后处理
postprecipitation 后沉淀
potable 适于饮用的
potable water 饮用水
potassium 钾
potassium bicarbonate 碳酸氢钾
potassium bisulfate 硫酸氢钾
potassium bisulfite 亚硫酸氢钾
potassium bromate method 溴酸钾法
potassium carbonate 碳酸钾
potassium chloride 氯化钾
potassium chromate 铬酸钾
potassium dichromate 重铬酸钾
potassium dichromate method 重铬酸钾法
potassium dihydrogenphosphate 磷酸二氢钾
potassium ferricyanide 铁氰化钾
potassium ferrocyanide 亚铁氰化钾
potassium hydroxide 氢氧化钾
potassium nitrate 硝酸钾
potassium perchlorate 高氯酸钾
potassium permanganate 高锰酸钾
potassium permanganate method 高锰酸钾法
potential 电势,电位;潜在的,可能的,势的
potential divider 分压器
potential energy 位能,势能
potential gradient 位梯度,势梯度
potential measurement 电位测量
potential of zero charge 零电荷电位
potential step 电势阶跃
potential sweep 电势扫描
potential-pH diagram 电位-pH 图
potentiometric 电势测定的,电位计的
potentiometric selectivity coefficient 电势法选择性系数
potentiometric sensors 电位传感器
potentiometric stripping analysis 电位溶出分析
potentiometric titration 电位滴定法
potentiometry 电位法,电势测定法
potentiostat 恒电势仪
potentiostatic 恒电位,恒电势的
potentiostatic method 恒电位法
powder forming 粉末成形
powder metal forging 粉末锻造
powder metallurgy 粉末冶金
powder metallurgy friction material 粉末冶金摩擦材料
powder metallurgy high speed steel 粉末冶金高速钢
powder metallurgy porous material 粉末冶金多孔材料
powder metallurgy structural parts 粉末冶金结构零件
powder paint 粉末涂料
power consumption analysis 消耗功率分析
power synthesis 功率合成
power density 功率密度
power distribution equipment 配电装置
power distribution panel 配电盘
power spectral density 功率谱密度
pre-annealing 预先退火
pre-cleaning 预清理
pre-dip 预浸
pre-polarization current 前极化电流
pre-test 初探,预检
pre-wave 前波,吸附波
prealignment 预先调准,预对准,预先对准,预校正
preceding 在前的,前述的
preceding reaction 前置反应
prechlorination 预氯化,预先加氯法
precipitate 沉淀,沉淀物;使沉淀;沉淀
precipitated 沉淀的
precipitating 起沉淀作用的

precipitation 沉淀
precipitation forms 沉淀形式
precipitation hardening 析出硬化，析出强化，沉淀硬化
precipitation heat treatment 析出热处理
precipitation titration 沉淀滴定法
precision 精密度
precision forging 精密锻造
preplating 预镀
precleaning 预洗净
precoat 预涂层
predetermined voltage 预定电压
prefabricated 预制的，预制构件的
preferential 优先的，选择的，先取的
preferential deposition 优先沉积［析出］作用
preferred 首选的，优先的
preferred orientation 择优取向
prefilming agent 预膜剂
preheating 预热；预热的
preliminary 初步的，预备的
preparation 制备，预备
prepared 制备的，精制的
preparing 准备的
prepolymer 预聚物
prepreg 预浸料坯，预浸材料
prepurge 预吹扫，预扫风
press forging 冲锻
press hardening 模压淬火
press quenching 加压硬化，压力淬火
press specification 冲床规格
press steel ball 防爆阀
pressing die 压模
pressing forming 挤出成型，挤塑
pressure 压力，压迫；密封，使……增压
pressure difference 压力差
pressure feed paint container 油漆增压箱
pressure feed type spray gun 压送式喷枪
pressure filter 压力式过滤器
pressure instrument 压力仪表/压
pressure mark 压痕

pressure meter 压力表
pressure probe 压力探头
pressure sintering 加压烧结
prestressed concrete 预应力混凝土
pressure tank 压力锅
pressure test 压力试验，试压
pressure vessel 压力容器
pressure evacuation test 压力抽真空试验
pretreatment 前处理；预处理期间的
pretreament chemicals 前处理剂
primary 主要的，初级的，基本的
primary batterie 原电池
primary coat 底漆，初涂，底涂层
primary coil 一次线圈，初级线圈
primary current distribution 初次电流分布
primary degradation 初级降解
primary phase 初生相
primary radiation 初级辐射，一次辐射，原辐射
primary reference electrode 一级参比电极
primary sedimentation tank 初次沉淀池
primary standard 基准物质
primary treatment 一级处理，初级处理
primer 底漆
primer coating 涂底漆，初级涂烘
primer surfacer 底漆二道浆
principal 主要的
principle 原理
printed circuit board 印刷电路板
printing ink 印刷油墨
probe 探针，调查；调查，探测
probe needle 探测针
probe test 探测试验，探头试验
process 过程，方法，步骤；经过特殊加工（或处理）的；处理
process annealing 临界温度以下退火，中间退火，制程退火，工序间退火
process pipe 工艺管道
process properties 工艺性能，工序性质
product 产物，产品
production plan control 生产计划控制

proeutectoid 先共析体，前共析
proeutectoid cementite 先共析渗碳体
proeutectoid ferrite 先共析铁素体
profile 剖面、侧面，外形，轮廓
profile steel 型钢
proof fabric 胶布
programmable 可设计的，可编程的
programmed 程序化的，程控的
programmed current chronopotentiometry 程控电流计时电势法
progressive 进步的，先进的
progressive bending 连续弯曲加工
progressive blanking 连续下料加工
progressive die 连续模，顺序冲模
projected 投影的
projected area 投影面积
property 性能，性质
propionic 丙酸的
propionic acid 丙酸
proportion 比例，部分，面积，均衡；使成比例，使均衡
proportional 比例的，成比例的，相称的，均衡的
proportional limit 比例极限
proportionality 相称，均衡，比例性
proportionality constant 比例常数
propoxylation 丙氧基化
propyl 丙基，丙烷基
protective 防护的
protective material 防护材料，保护剂
protective potential 保护电位
protonation 质子化作用，加质子作用
proton 质子
proton-transfer reactions 质子转移反应
protonic 质子的
protonic solvent 质子溶剂
prototropic 质子异变的
prototropic rearrangement 质子转移重排
prototropy 质子转移
proximate 近似的，最近的
proximate analysis 近似分析

proximity 接近，邻近
proximity bake 邻近烘烤处理
proximity effect 邻近效应
pseudo 冒充的，假的
pseudo-inductance 假电感
pull strength 拉引强度，拉力强度，拉拔强度
pull test 拉引试验，拉伸试验，张力试验
pulsation 脉动，搏动，震动
pulse 脉冲；使跳动
pulse amplitude 脉冲幅度
pulse echo method 脉冲回波法，脉冲反射法
pulse energy 脉冲能量
pulse length 脉冲长度
pulse plating 脉冲电镀
pulse polarography 脉冲极谱法
pulse repetition frequency 脉冲重复频率
pulse repetition rate 脉冲重复率
pulse reverse 反向脉冲
pulse tuning 脉冲调谐
pulse voltammetry 脉冲伏安
pulsed amperometric detection (PAD) 脉冲电流检测
pulsed pulse 脉动脉冲
pump 泵；打气，用抽水机抽……
pump-out time 抽气时间
pump-out tubulation 抽气管道
punch 冲头
punch press 冲床
punch riveting 冲压铆合
punch-cum-blanking die 凹凸模，高低模
punched hole 冲孔，穿孔
purge 净化
putty 腻子，油灰
pycometer 比重瓶
pyknowmeter flasks 容量瓶
pyranose 吡喃糖
pyridine 吡啶
pyrolysis 高温分解，热解
pyrolytic coating 热解涂层，高温喷涂

pyrolytic decomposition 热解，裂解，高温分解
pyrolytic elimination 热解消除
pyrolytic graphite 热解石墨
pyrophosphate 焦磷酸盐，焦磷酸酯
pyrophosphate copper 焦铜
pyroplastic deformation 高温塑性变形
pyrrolidone 吡咯烷酮

Q

quad 四边形，方形，扇形体，四倍的，由四部分组成的

quadr- 【构词成分】四，平方，二次

quadrangle 四边形，方形；四边形的，方形的

quadrangular 四边形的，四角形的

quadrantal 象限的，四分圆的，四分仪的

quadrate 正方形，长方骨；正方形的，方形的，使成正方形，使适合

quadratic 平方的，二次的，（正）方形的

quadratic crystal 正方晶体

quadratic form 二次形式

quadrature 求积，求积分，正交

quadravalence 四价

quadravalency 四价

quadravalent 四价的

quadri- 【构词成分】四，四倍，四重

quadribasic acid 四价酸，四元酸

quadric- 【构词成分】二次的

quadricovalent 四配价的

quadrimolecular reaction 四分子反应

quadripolymer 四单体共聚物，四元聚合物，四元共聚物

quadrivalence 四价

quadroxide 四氧化物

quadruple 四倍；四倍的，四重的，使……成四倍

quadruple effect evaporator 四效蒸发器

quadrupole mass spectrometer 四极杆质谱仪，四极质谱仪

quadrupole moment 四极矩

quadrupole resonance 四极共振

quadrupling 四倍；使成四倍

qualification 资格，条件，限制，限定，赋予资格

qualificatory 限定的，赋予资格的，使合格的，限制性的，带有条件的

qualified 合格的，有资格的，经过鉴定的

qualified products 合格品，良品

qualify 取得资格，有资格；限制，使具有资格，证明……合格

qualimeter X 射线（穿透）硬度计

qualitative 定性的，质的，性质上的

qualitative analysis 定性分析

qualitative analysis of organic functional group 有机官能团定性分析

qualitative filter paper 定性滤纸

qualitative reaction 定性反应，定性检验

quality 品质，质量

quality analysis 质量分析

quality assurance 质量保证，品质保证

quality control 质量控制，质量管理

quality improvement 品质改善，质量改进

quality sign 品质标记

quality test 质量试验，质量检验

quantification 定量，量化

quantify 量化，为……定量，确定数量；量化，定量

quantitate 测定的数量，估计的数量，用数量来表示

quantitative 定量的，量的，数量的

quantitative adsorption 定量吸附

quantitative analysis 定量分析

quantitative classification 定量分类

quantitative filter paper 定量滤纸

quantitative volumetric analysis 定量容积分析

quantity 量，数量，大量，总量

quantity factor 定量参数

quantity of heat 热量

quantity of state 状态量

quantity production 大量生产，批量生产

quantivalence 原子价，化合价

quantivalency 化合价

quantivalent 多价的

quantization 量子化，分层，数字化，

量化
quantize 使量子化，数字转换
quantometer 光量计
quantum 量子论，量子，数量，和，总量，时限
quantum chemistry 量子化学
quantum condition 量子条件
quantum efficiency 量子效率
quantum hypothesis 量子假说
quantum jump 量子跃迁
quantum leakage 量子漏泄
quantum liquid 量子液体
quantum mechanics 量子力学
quantum number 量子数
quantum orbit 量子轨道
quantum path 量子轨道
quantum scattering 量子散射
quantum state 量子状态
quantum theory of valence 价的量子理论
quantum theory 量子论
quantum yield 量子效率
quaquaversal 穹形的
quark 夸克（理论上一种比原子更小的基本粒子）
quark model 夸克模型
quarry-faced 粗面的，毛面的
quartation 四分法
quarter 四分之一；把……四等分
quartern 四分之一，四等分
quarternary 四元的，四进制的，四级的
quartet 四重峰
quartz 石英
quartz crystal microbalance（QCM） 石英晶体微天平
quartz glass 石英玻璃
quartz lens 石英透镜
quartz plate 石英片
quartz sand 石英砂
quartz spectrograph 石英摄谱仪
quartz tube 石英管
quartzite 石英岩

quasi 准的，类似的，外表的；似是，有如
quasi-continuous 准连续的
quasicleavage 准解理断裂
quasicrystal 准晶体
quasi elastic force 准弹性力
quasilinearization 拟线性化
quasi-isotropy 类无向性
quasi-orthogonal 准正交线；拟正交的
quasiracemate 准外消旋体
quasi-regularity 准正则性，拟正则的
quasi-reversible 准可逆的
quasi saturation 准饱和
quasi static process 准静态过程
quasi stationary state 准稳状态
quasi-synchronization 准同步
quaternary 四，四个一组；四进制的，四个一组的
quaternary ammonium compound 季铵化合物
quaternary ammonium base 季铵碱
quaternary ammonium salt 季铵盐
quaternary base 季碱
quaternary carbon atom 季碳原子
quench 熄灭，淬火，解渴，结束，冷浸，急冷
quench aging 淬火老化
quench oil 淬火油
quenchant 骤冷剂，淬火剂，淬火介质
quenched charcoal 淬火炭
quenching 熄灭，淬火，骤冷
quenching and tempering 调质处理，淬火及回火
quenching bach 淬火槽
quenching crack 淬火裂痕
quenching distortion 淬火变形
quenching stress 淬火应力
queue 队列，长队；使……排队
quick analysis 快速分析
quick charge 快速放电
quick drying ink 快干墨水

quick fermentation	快速发酵
quick setting	快凝
quicking	汞齐化
quicklime	氧化钙
quicksilver	汞
quiescent	静止的,不活动的,沉寂的
quinary	五个一套或一组;第五位的
quinazoline	喹唑啉
quincuncial	五点梅花状排列的,五点形的
quinhydrone	氢醌
quinhydrone electrode	氢醌电极
quinic acid	奎宁酸
quinine	奎宁
quinoid structure	醌型结构
quinoid	醌式
quinoline	喹啉,喹啉衍生物
quinoline dye	喹啉染料
quinolinic acid	喹啉酸
quinolylamine	氨基喹啉
quinone	醌
quinoneimine dye	醌亚胺染料
quinoxaline	喹喔啉
quinquangular	有五个角的,五角形的
quinquevalence	五价
quinquevalency	五价
quintet	五重峰
quintuple	五倍量,五个一组;五倍的,五部分组成的;使成五倍
quota	配额,定额,限额
quotation	引证,引语
quote	引用;报价,引用,引证,举证
quotient	商,系数,份额
quotiety	系数,率
quoting	引号,引用

R

racemate 外消旋化合物
racemic 外消旋的，消旋酸的
racemic acid 外消旋酸
racemic modification 外消旋体，外消旋变体
racemism 外消旋性
racemization 外消旋
rack plating 挂镀
racking 上挂具
radial sectioning 径向剖面，径向切片
radiance 辐射，光辉，发光
radiance temperature 发光温度，辐射温度
radiant 辐射的
radiant arc furnace 辐射电弧炉
radiant energy 辐射能
radiant flux 辐射流
radiant heat 辐射热
radiant ray intercepted glass 辐射线防御玻璃
radiation 辐射
radiation aging 辐照老化
radiation chemical reaction 辐射化学反应
radiation chemistry 辐射线化学
radiation cure 辐射硫化
radiation curing 辐射干燥
radiation damage 辐照损伤
radiation dose 辐射剂量
radiation equilibrium 辐射平衡
radiation hazard 放射危害
radiation intensity 辐射强度
radiation law 辐射律
radiation loss 辐射损失
radiation polymerization 辐照聚合
radiation protection 辐射防护
radiation pyrometer 辐射高温计，辐射高温表
radiation shield 辐射屏蔽
radiation temperature 辐射温度
radiationless transition 无辐射跃迁
radiative recombination 辐射复合
radiative transfer 辐射传热
radical 基
radical ion 自由基离子
radical polymerization 游离基聚合
radical scavenger 游离基清除剂
radio frequency 射频
radio frequency heating 高频加热
radioactive 放射性的，有辐射的
radioactive contamination 放射性污染
radioactive decay 放射性衰变
radioactive deposit 放射性沉积物
radioactive disintegration 放射性衰变，放射性蜕变
radioactive element 放射性元素
radioactive equilibrium 放射平衡
radioactive indicator 放射指示剂
radioactive isotope 放射性同位素
radioactive paint 放射性油漆
radioactive substance 放射性物质
radioactive tracer 放射指示剂
radioactive transformation 放射性转换
radioactive waste 放射性废物
radioactivity 放射能力，放射性
radiocarbon 放射性碳
radiochemical analysis 放射化学分析
radiochemistry 放射化学
radiochromatogram 辐射色层分离谱，辐射色谱图
radiochromometer 放射比色计
radiocolloid 放射性胶体
radioelectrochemistry 放射电化学
radioelement 放射性元素
radiograph 射线照相
radiographic 射线照相术的
radiographic inspection 放射线探伤法，射线照相检验

radiographic test	射线检验，放射线探伤
radiography	射线照相术
radioisotope	放射性同位素
radiology	射线学，放射学
radiolysis	辐照分解
radiometer	辐射计
radiometric	放射性测量的，辐射度的，辐射测量的
radiometric analysis	放射分析法，辐射度分析
radiometric titration	放射滴定
radiometry	辐射测量术
radionuclide	放射性核素
radiosensitivity	放射敏感性
radiotracer	放射指示剂，放射性示踪剂
radium	镭
radius	半径
radius ratio	半径比
raffinate	提余液，剩余液
raffinose	棉子糖
rag paper	粗纸，棉浆纸
Raman effect	拉曼效应
Raman spectrum	拉曼光谱
rancidity	酸败
random error	偶然误差，随机误差
random variable	随机变数
range	极差，量程，范围，射程
range of explosion	爆炸范围
rapid analysis	快速分析
rapid cooling system	急冷系统
rapid cure adhesives	速定胶粘剂
rapid thermal process	快速热处理
rare earth elements	稀土元素
rare gas	稀有气体
rare metal	稀有金属
rarefaction	稀薄化，稀疏
raster	光栅
rate constant	速度常数
rate current	反应电流
rate determining step	速率控制步骤
rated capacity	额定容量
rated voltage	额定电压
rate-determining steps	速率决定步骤
ratio	比率
rational	合理的，理性的
rational curve	特性曲线
rattler	滚筒，打磨滚筒
rattler test	磨耗试验
raw data	原始数据
raw material	原料
raw rubber	生橡胶
raw sewage	原污水
raw silk dyeing	生丝染色
raw vegetable oil	生植物油
raw water	原水
reactance	电抗
reactant	反应物，反应剂，试剂
reaction	反应
reaction accelerator	反应加速剂
reaction chain	反应连锁
reaction chamber	反应室
reaction coordinate	反应坐标
reaction current	反应电流
reaction formula	反应式
reaction intermediate	反应中间体
reaction isochore	反应等容线
reaction isotherm	反应等温线
reaction kinetics	反应动力学
reaction mechanism	反应机理，反应历程
reaction mixture	反应混合物
reaction order	反应级数
reaction product	反应产物
reaction promotor	反应促进剂
reaction rate	反应速率
reaction velocity constant	反应速度常数
reaction vessel	反应器
reaction zone	反应带
reactivation	再活化，重激活
reactive	反应的
reactive dye	活性染料

reactive intermediate	活泼中间体
reactive ion beam etching system	反应离子束蚀刻系统
reactive sputter etching system	反应性溅镀蚀刻系统
reactive sputtering system	反应性溅镀系统
reactive vessel	反应器
reactivity	反应性
readily biodegradable substances	易生物降解物质
ready mixed paint	低漆，调和漆
reagent	反应物，试剂
reaming	铰孔加工，铰孔修润
reboil	再沸
reboiler	再沸器
recalibration	复校，重新校准
receptacle	插座
receptor	接受体
recharge	再充电
rechargeable	可再充电的
rechargeable batteries	充电电池
reciprocal	相互的，倒数的，彼此相反的
reciprocal lattice	倒易晶格
reciprocal proportion	反比例
reciprocal reaction	可逆反应
reciprocation	交换，往复运动
reciprocation pump	往复泵
reciprocity law	倒易律，互易定律，互反律
recirculating cooling water	循环冷却水
recirculation	再循环
reclaim	再生
reclaimed rubber	再生胶
recoatability	重涂性
recoil	弹回，反作用
recoil energy	反冲能
recombination	再结合，再调质，复合
recombination coefficient	复合系数
recooler	二次冷却器
recorder	记录仪
recording hygrometer	湿度计
recording manometer	自记压力计
recording thermometer	自记温度计，温度记录器
recovering charge	恢复充电
recovery	回收，恢复
recovery installation	回收设备
regeneration	再生
recrystallization	再结晶
rectification	精馏
rectifier	整流器
rectocondensor	直型冷凝器
recycle	再循环，再生
recycle mixing	循环混合
recycle ratio	循环比
recycle stock	再循环物料
recycling	再回收，回收利用
red copper	紫铜
red copper ore	赤铜矿
red heat	赤热
red iron ore	赤铁矿
red iron oxide	铁红
red lead	红丹
red mercuric oxide	氧化汞
red mud	红泥
red prussiate	铁氰化钾
red shift	红移
red shortness	红热脆性
redistillation	再蒸馏
redox	氧化还原反应
redox catalysis	氧化还原催化
redox catalyst	氧化还原催化剂
redox couple	氧化还原电对
redox indicator	氧化还原指示剂
redox polymerization	氧化还原聚合
redox potential	氧化-还原电位
redox potentiometry	氧化还原电位滴定
redox reaction	氧化还原反应
redox system	氧化还原系统
redox titration	氧化还原滴定
reduced	减少的，还原的
reduced iron	还原铁

reduced pressure distillation	真空蒸馏
reduced pressure method	减压法
reducer	还原剂
reducidility	稀释剂
reducibility test	稀释试验
reducing agent	还原剂
reducing roasting	还原焙烧
reducing sugar	还原糖
reducing valve	减压阀
reductant	还原剂
reductase	还原酶
reduction	还原，减少
reduction current	还原电流
reduction potential	还原电势［位］
reduction wave	还原波
reduction zone	还原区
reductive acylation	还原酰化
reductive alkylation	还原烷基化
reductive dimerization	还原二聚
reductometry	还原滴定法
reductor	脱氧剂，还原剂
reel stretch	卷圆压平
reference	参考文献，参考，基准
reference drawing	参考图
reference electrode	参比电极
reference fuel	参考燃料
Reference hydrogen electrode	参比氢电极
reference material	参考材料
reference solution	参考溶液
reference standard	参考标准
refining	精制，净化
refining mill	精磨机
reflectance	反射系数
reflection	反射
reflection coefficient	反射系数
reflection density	反射密度
reflection factor	反射系数
reflectivity	反射率，反射性
reflector	反射器反射镜，反射体
reflux	回流
reforming	改造，重整
refraction	折射
refractive index	折射率
refractometer	折射计
refractories	耐火材料
refractoriness	耐火度，耐火性
refractory	耐火的
refractory alloy	耐热合金
refractory coating	耐火涂料
refractory hard metal	耐火硬金属
refrigerant	冷冻剂
refrigerating capacity	制冷能力
refrigerating industry	制冷工业
refrigeration	冷却
refrigeration condenser	制冷凝汽器
refrigeration cycle	冷却循环
regenerant	再生剂
regenerated	再生的
regenerated cellulose	再生纤维素
regenerated fiber	再生纤维
regeneration	再生
regenerative furnace	回热炉
regenerator	回热器
region	部位
registration	登记
regrinding	再次研磨
regular polymer	规则聚合物
regular solution	正规溶液
regulation	整顿；规定的，平常的
reinforced	加固的，加强的，加筋的
reinforced fiber separator	强化纤维隔板
reinforced material	增强材料
reinforced plastics	增强塑料
reinforcement	增强
reinforcement measure	加固措施
reinforcing agent	增强剂
reinforcing filler	增强填料
reinforcing pad	补强垫，增强衬板
reinforcing pigment	加强颜料
retentivity	顽磁性
reject	次品

residual affinity 剩余亲和势，残余亲和力
relative 相对的，有关系的，成比例的
relative abundance 相对丰度
relative aperture 相对孔径
relative area 相对面积
relative average deviation 相对平均偏差
relative error 相对误差
relative humidity 相对湿度
relative magnetic permeability 相对磁导率
relative permeability 相对磁导率
relative resolution of a spectrometer 谱仪相对分辨率
relative spectral distribution 分光分布
relative standard deviation 相对标准偏差
relative velocity 相对速度
relative viscosity 相对黏度
relative volatility 相对挥发度
relativity 相对性
relaxation 松弛，松弛极化
relaxation mechanism 弛豫历程
relaxation modulus 松弛模量
relaxation phenomena 松弛现象
relaxation time 松弛时间
release agent 脱离剂，脱模剂，隔离剂
relevant indication 相关指示
relief valve 安全阀
reluctance 磁阻
remedy 补救
remelting 再熔
remote control 遥控
render （焊接，油漆的）打底
reorganization energy 重组能
repainting 重新涂装
repair welding 补焊
repetition 重复，复制品
replacement 置换
replacement reaction 置换反应
replacement titration 置换滴定
reprecipitation 再沉淀
representative 典型的，有代表性的
reproducibility 再现性

reproduction 再生产，再生，复制品
requirement 要求
reservoir 水库，蓄水池
reset 复位
residence time 停留时间
residual 剩余，残渣；剩余的，残留的
residual capacity 残存容量
residual chlorine 残留氯
residual current 残余电流，剩余电流
residual energy 剩余能
residual hardness 剩余硬度
residual magnetic field 剩磁场
residual magnetic technique 剩磁技术，剩磁法
residual magnetism 剩磁
residual nitrogen 剩余氮
residual oil 残油
residual rays 剩余射线
residual resistance 剩余电阻
residual stress 残留应力，残余应力
residual sulphur 残留硫
residue 残留物，滤渣
resin streak 树脂流纹
resin trap 树脂捕捉器，树脂收集器
resin wear 树脂脱落
resinoid 热固性黏合剂，树脂状物质，树脂状的
resist 抗蚀剂，防染剂
resist printing 防染印花
resist processing equipment 抗蚀剂处理设备
resist stripper 抗蚀剂剥离液
resist stripping system 抗蚀剂剥离系统
resist temperature control 抗蚀剂温度控制
resist thermal stability 抗蚀剂热稳定性
resist thickness uniformity 抗蚀剂膜厚均质性
resistance 电阻，阻力，阻抗
resistance furnace 电阻炉
resistance heating 电阻加热

resistance heating evaporation system 电阻加热真空蒸镀系统
resistance pyrometer 电阻高温表
resistance thermometer 电阻温度计
resistivity 电阻率，电阻系数
resolution 分离度，分［辨］解，离析，溶解
resonance 共［振］鸣，谐振
resonance absorption 共振吸收
resonance energy 共振能
resonance hybrid 共振杂化
resonance method 共振法，谐振法
resorption 再吸收
respective 分别的，各自的
respectively 分别地，各自地，独自地
response factor 响应系数，响应因子
restriking 二次精冲加工，再闪击，电弧重燃，矫形锻压
retained austenite 残留奥氏体
retardant paint 防火油漆，酚醛防火漆
retardation 减速，阻滞，迟延
retardation phenomenon 延滞现象，滞后现象
retardation time 滞后时间
retarder 防潮剂，缓凝剂，阻滞剂
retarding agent 阻滞剂，抑制剂
retarding field analyser 减速场分析器
retention 保留
retention time 保留时间
retention volume 保留容积
reticular 网状的，错综的
reticular structure 网状结构，网状组织
reticulated vitreous carbon electrode 网状玻璃碳电极
retrogradation 降解
returned activated sludge 回流活性污泥
reuse 再利用
reutilization 再用，二次利用，重复利用
reverberatory 反射炉，反应炉；反射的，反响的
reverberatory furnace 反射炉

reversal 逆转，反转，相反的
reversal development 反转现象，反转显影法
reversal effect 反转效应
reversal process 反转成像法
reversal technique 反向技术
reverse 背面，相反；反面的；颠倒，倒转
reverse angle 倒角
reverse charge 反充电，反向充电
reverse current 逆电流
reverse osmosis 反渗透
reverse pulse voltammetry 反向脉冲伏安法
reverse reaction 逆反应
reverse side 反面，背面
reversed 相反的，颠倒的
reverser clean 转动式清洁器
reversibility 可逆性
reversible 可逆的
reversible cell 可逆电池
reversible change 可逆变化
reversible colloid 可逆胶质
reversible electrode 可逆电极
reversible gel 可逆凝胶
reversible hydrolysis 可逆水解
reversible process 可逆过程
reversible reaction 可逆反应
reversible temper brittleness 可逆回火脆性，第二类回火脆性，高温回火脆性
reversible wave 可逆波
reversing 回归
revolver gress 旋压机
revolving furnace 旋转炉
revolving screen 旋转筛
reynolds number 雷诺数
rhenium 铼
rheological 流变学的，流变性，流变的
rheological hysteresis 流变滞后
rheology 流变学，液流学
rheology control 流变性控性
rheometer 流变仪
rheostat 变阻器

rheopexy 震凝现象，触变性，抗流变性
rhodanide 硫氰化物，硫氰酸盐
rhodanizing 镀铑
rhodanometry 硫氰酸盐滴定法
rhodate 铑酸盐
rhodium 铑
rhodium plating process 铑（白金）电镀工艺
rhombic 菱形的，斜方形的，正交的
rhombic sulfur 斜方硫
rhombic system 正交晶系
rhyotaxitic 流纹状的，流纹状
rider 游码
rifle color coating 枪色镀层
rigid 刚性的，硬性的
rigid body 刚体
rigid foam 硬质泡沫塑料，硬塑胶
rigid foam rubber 硬泡沫橡胶
rigid matrix electrode 刚性基质电极
rigid plastic 硬质塑料
rigid polyvinyl chloride 硬聚氯乙烯
rigidity 刚性
rigidize 硬化，加固
rigidly 刚性地
rim 边，边缘，圆圈；作……的边，装边于……
rimmed steel 沸腾钢
ring 环形物；成环形
ring cathode 环状阴极
ring current 环流，回路电流
rinse 清洗，水洗
rinsing 清水，残渣；（用清水）冲洗
ripple 波纹；在……上形成波痕
rive 撕开；裂开，破裂
rivelling 条纹；弄皱，起皱，（使）皱缩
rivet 铆钉；铆接，固定
riveting die 铆合模
roaster 焙烧炉
roasting 焙烧
roasting furnace 焙烧炉

roasting kiln 燃烧炉
robot 机械手，机器人
robustness 强度，坚固性，耐久性
rocking die forging 摇动锻造
rockweel hardness 洛氏硬度
rockweel hardness test 洛氏硬度试验
roll crusher 滚碎机，破碎机
roll bending 滚筒弯曲加工
roll material 卷料
roll mill 滚碎机，辊式捏合机
roller 滚筒
roller coating 滚涂
roller hearth oven 滚式炉
rolling 旋转；旋转的，起伏的，波动的
rolling machine 碾压机
roof angle 顶角
room temperature 室温
room temperature curing method 室温固化法
room temperature setting adhesive 室温固化黏合剂
ropiness 黏着性，成黏性的丝
ropy 绳状的，黏稠的，可拉成丝的
rosin 松香，树脂
rosin oil 松香油
rosin soap 松香皂
rosin spirit 松香精
rosin-based polyglucoside 松香基葡萄糖苷
rot 腐烂，腐败；使腐烂
rot (a) - 【构词成分】（旋）转
rotamer 旋转异构体
rotary 旋转的，转动的
rotary air sander 转动型气动打磨机
rotary atomizing electrostatic spraying equipment 旋杯式静电喷涂装置
rotary coupon test 旋转挂片试验
rotary crusher 旋转压碎机
rotary dryer 旋转干燥器
rotary filter 旋转滤器
rotary forging 回转锻造
rotary pump 旋转泵

rotated dropping mercury electrode	旋转滴汞电极
rotating crystal method	旋转晶体法
rotating disk electrode（RDE）	旋转圆盘电极
rotating ring disk electrode	旋转环盘电极
rotation	旋转，循环，轮流
rotation axis	旋转轴
rotation vibration spectrum	转动振动谱
rotational	转动的，回转的
rotational energy	转动能
rotational isomer	旋转异构体
rotational isomerism	旋光异构
rotational level	转动能级
rotational magnetic field	旋转磁场
rotational molding	离心成形，旋转模塑，旋转成型
rotational scan	转动扫查，旋转扫查，旋转扫描
rotational viscometer	旋转黏度计
rotoblast	转筒喷砂，喷丸
rotogravure	印刷用滚筒
rotproofness	防腐性，耐腐性
rottenness	脆性，易碎性，腐烂
rough	粗糙的，未经加工的
rough machining	粗切削，粗加工
rough sand	粗砂
rough sketch	草图，简略图
roughening	粗化
roughness	粗糙度，毛刺
round corner	圆角
round flask	圆底烧瓶
round-bottom flask	圆颈烧瓶
rounding	圆形加工；圆的，使圆的
routine analysis	常规分析
rub	摩擦，障碍，磨损处；擦，摩擦
rubber	橡胶
rubber dough	橡胶胶水
rubber fabric	胶布，涂胶带
rubber insulation	橡胶绝缘
rubber latex	胶乳
rubber molding	橡胶成形
rubber patten	橡胶磨块
rubber solvent	橡胶溶剂
rubber sponge	海绵橡胶
rubber state	橡胶态
rubber stopper	橡皮，橡胶塞
rubber substitute	油膏
rubber suction bulb	洗耳球
rubber thread	橡胶线
rubber toughening	橡胶增韧
rubber tube	橡皮管
rubbing oil	摩擦用油
rubbing	磨损，摩擦，研磨
rubidium	铷
rude	天然的，原始的，加工粗糙的
ruggedise	使坚固，使耐用，增加耐磨和可靠性
rugosity	皱纹，凸凹不平，粗糙度，不规则
rumbler	清理滚筒
running	流淌；连续的，流动的，运转着的
rupture	破裂；使破裂
rust	锈，生锈；使生锈，腐蚀
rust inhibiting paint	防锈涂料
rust inhibitor	抗腐蚀添加剂
rust prevention	防蚀，防蚀处理
rust preventing pigment	防锈颜料
rust preventive	防锈剂
rust preventive oil	防锈油
rust proof paint	防锈涂料，防锈漆
rust spotting	锈斑
ruthenium	钌
rutile	金红石

S

saccharide 糖类，糖
secchariferous 含糖的，产生糖的
saccharin 糖精，邻磺酰苯甲酰亚胺
saccharose 蔗糖
sacrificial 牺牲的
sacrificial anode 牺牲阳极
sacrificial corrosion 牺牲（阳极）腐蚀
safe handling 安全操作
safe operating area 安全运转区域，安全工作区
safety device 安全装置，保护装置，过载安全装置
safety engineering 安全工程
safety explosive 安全炸药
safety factor 安全系数
safety fuel 安全燃料
safety valve 安全阀
sag 下垂，流挂；使下垂
sag curve 挠度曲线，垂度曲线
sagging 松垂，流淌
sal ammoniac 卤砂，氯化铵
sal mirabile 硫酸钠，芒硝
sal soda 碳酸钠，苏打
sal volatile 碳酸铵，挥发盐
salicyl 水杨基，邻羟苯基
salicyl alcohol 水杨醇，邻羟苯甲醇
salicyl aldehyde 水杨醛
salicylamide 水杨酰胺
salicylate 水杨酸盐
salicylic acid 水杨酸
salifiable 能与盐混合（或化合）的，能成盐的
salify 使成盐，使与盐结合
saligenin 水杨醇
salimeter 盐液密度计，含盐量测定计
salimetry 盐度测量法
salination 用盐处理，盐化
saline 盐水；盐的，含盐分的

saline concentration 含盐量
saline matter 盐分
salineness 含盐度
saliniferous 含盐的，盐土的
salinity 盐度，盐分，盐浓度，含盐量
salinization 盐化作用，盐碱化
salinous 盐的，咸的
saliter 硝酸钠，硝石
salicylic acid 水杨酸
salmiac 氯化铵，硇砂，卤砂
salometer 盐量计，盐液比重计，盐液密度计
salometry 盐浓度测量法
salt 盐
salt bath 盐浴
salt bath quenching 盐浴淬火
salt bridge 盐桥
salt effect 盐效应
salt elimination 除盐
salt error 盐误差
salt shrinking 盐缩
salt solutions 盐溶液
salt spray cabinet 盐雾室
salt spray test 盐雾试验
salt tolerance 耐盐性，耐盐度
salt water resistance 耐盐水性
salt removal 脱盐的
salted 盐的，用盐处理的
salting out 盐析
salting out effect 盐析效应
saltness 含盐度，咸性
salty 咸的，含盐的
salvage （废物）利用，废弃品（处理），废物处理，（可利用的）废料
salvage department 废料（利用）车间
salvage point 废品收集处
salvage sump 废油收集池
salvaged material 废弃物

samaria 氧化钐
samaric 三价钐的
samaric chloride 氯化钐
samarium 钐
samarous （亚，二价）钐的，亚钐化物
samarous chloride 二氯化钐
samite 碳化硅，金刚砂
sample 样品
sample handling 样品处理
sample solution 样品溶液
sampled current voltammetry 取样电流伏安法
sampler 取样器，采样器
sampling 取样，抽样
sampling inspection 抽样［取样］检查
sampling method 取样方法
sampling probe 取样［采样］探头
sampling probe function 取样探测功能
sand aperture 砂眼
sand bath 砂浴
sand blast (ing) 喷砂
sand cloth 砂布
sand filter 砂滤器
sand filtration 砂滤
sand lime brick 石灰砂粒碳，灰砂砖
sand mold casting 砂模铸造
sand paper 砂纸
sand scratch 打磨划伤
sander 砂光机，研磨机，砂磨机
sanding agent 研磨剂
sanding cloth 打磨砂布
sanding disc 金刚砂研磨盘
sanding scratch 打磨器纹，砂纸纹
sanding sealer 二度底漆，掺砂涂料
sandish 砂（质）的
sandslinger 抛砂机，投砂器，抛砂造型机
sandwich cells 夹层电池
sanforizing （织物）防缩处理，机械预缩整理
sanshoamide 山椒酰胺
sanshool 山椒醇

santomerse 润湿剂，烷化芳基磺酸盐
saponaceous 肥皂似的，肥皂质的
saponated 经皂处理过的，皂化的
saponid 合成洗涤剂
saponifiable 可皂化的
saponifiability 皂化性，可皂化性
saponification 皂化
saponification value 皂化值
saponified 皂化过的
saponifier 皂化剂
saponify 皂化，使皂化
saponifying agent 皂化剂
saponin 皂荚苷，皂素
sapphire 蓝宝石，天蓝色；天蓝色的
sapphire whisker 氧化铝纤维
saprobia 污水生物，腐生生物
saprobic 腐生的，污水生物的
saprobiont 污水生物，腐生物
saprogenic 生［产］腐的，腐化的
saprolite 残余土，腐泥土
saprophage 腐生物
sarcosine 肌氨酸，N-甲基甘氨酸
sark 衬垫物
sarking 衬垫材料，衬垫层
sassoline 天然硼酸
satin 缎子；光滑的，绸缎做的，似缎的
satin finish 缎面加工，光泽装饰，施釉，磨光
satin nickel 尼龙镍，砂镍，雾镍
satin paper 蜡光纸，光泽纸
satin texture 缎面咬花
satin tin 砂锡，缎锡
satin white 缎白
saturability 饱和度，饱和能力
saturable 可饱和的，能浸透的
saturate 浸透的，饱和的，深颜色的；浸透，使湿透，使饱和，
saturated 饱和的，渗透的，深颜色的
saturated acid 饱和酸
saturated activity 饱和活性
saturated calomel electrode 饱和甘汞电极

saturated compound 饱和化合物
saturated extent of adsorption 饱和吸附量
saturated fatty acid 饱和脂肪酸
saturated hydrocarbon 饱和烃
saturated liquid 饱和液体
saturated solution 饱和溶液
saturated steam 饱和蒸汽
saturated vapor pressure 饱和蒸汽压
saturating 饱和的，浸透的
saturation 饱和，饱和度，彩色度（鲜艳度、饱和度或纯度）
saturation curve 饱和曲线
saturation magnetic 磁饱和
saturation point 饱和点
saturation temperature 饱和温度
saturite 饱和溶液沉积物
saturnine 铅中毒的
saturnism 铅中毒，铅毒
saw 锯片
sawing 锯削［切］
sawing machines 锯床
saxin 糖精
saxoline 液体石蜡油
saybolt viscometer 赛氏黏度计
scab corrosion test 疤形腐蚀试验
scalar 标量，数量；标量［数量，梯状，分等级］的
scalariform 阶纹的，梯状的
scalary 如梯的，有阶段的
scale 结垢，刻度，氧化皮，标度
scale inhibitor 阻垢剂
scale pan 天平盘
scalehandling 清除氧化皮，消除水垢
scalelike 鳞状的
scaliness 有鳞，多鳞，起鳞程度
scaling 起（生成）氧化皮，起鳞，脱层，分层（缺陷），剥落
scalp 剥去表皮，刮［剥］光
scalping 剥皮，去表面层，修整，筛出粗块
scaly 有鳞的，积垢的，劣等的

scandium 钪
scanning 扫描；扫描的，观测的，搜索的
scanning electron microscope 扫描电子显微镜
scanning acoustic microscope 扫描声学显微镜
scanning auger microprobe (SAM) 扫描俄歇微探针
scanning electrochemical microscope (SECM) 扫描电化学显微镜
scanning electron microscope (SEM) 扫描电子显微镜
scanning path 扫描路径
scanning probe microscope (SPM) 扫描探针技术
scanning speed 扫描速度
scanning transmission electron microscope 扫描透射型电子显微镜
scanning tunneling microscope (STM) 扫描隧道显微镜
scar 伤痕；给留下伤痕，结疤
scarfjoint 嵌接
scatterance 散布［射］
scattered 分［疏，弥，扩］散的，散乱的
scattered light 散射光
scattered reflection 扩散反射，漫反射
scattered radiation 散射辐射
scattered spectrum 散射光谱
scattering 散射
scattering coefficient 散射系数
scattering cross section 散射截面
scattering light 散射光
scavenge 打扫，排除废气；清除污物，打扫
scavenger 净化剂
scent 气味；使充满……的气味
Schiff base 席夫碱
schillerization （晶体的）青铜光泽，青铜光泽现象
schistose 片岩的，片岩质的，片岩状的
schistose structure 片状构造，层状结构，

叶片状构造
schohartite 重晶石
schrauben 螺栓
scission 切断，分离，断开
sclero- 【构词成分】硬，硬化
sclerosal 硬的，硬化的
scleroscope 硬度计
scleroscopic 硬度计的
sclerose 硬化，使硬化
sclerometer 硬度计
scope 范围
scorch 烧焦，焦痕；焦化
score 刻［伤，划，截］痕，裂缝
scoria 熔［矿，炉，铁，金属，熔析］渣
scoriated 成熔渣的
scorification 烧熔，造渣，渣化
scoriform 熔渣状的，像熔渣的，似火山渣的
scorify 使变成矿渣，析取
scoring 刮伤，刻痕
scotch 刻痕；弄伤，制止转动
scotch tape 透明胶带
scour 擦，冲刷，洗涤剂；擦亮，洗涤，冲洗，清除
scouring 洗涤，擦亮，洗擦
scrack 划伤
scrap jam 废料阻塞
scrape 刮掉，擦痕，刮擦声；刮，擦伤，挖成
scrap 残余物，废料
scrap material 废料
scraper 刮刀
scrapings 刮料，被刮削下的碎屑
scrapless 无废料，无渣的
scrapless machining 无废料加工
scratch hardness 抗刮硬度
scratch resistant 耐刮擦度
scratch 刮，刮伤，划痕
scratchability 柔软度，易刻度，刻痕度
screen 屏，幕，屏风，筛子；筛
screen analysis 筛析，颗粒分析

screen mesh 丝网，筛孔，筛眼
screen pack 网组，过滤网版，过滤网组合
screen printing 筛网印花，丝网印刷
screen printing forme 丝网印版
screen printing frame 印刷网框
screenage 屏蔽，影像
screened 筛过的，屏蔽的
screening 遮蔽，掩护，隔离；筛分，筛选
screening agent 遮蔽剂，掩蔽剂
screening constant 屏蔽常数
screening effect 屏蔽效应
screening material 屏蔽材料
screening test 筛分试验
screw 螺丝
screw compressor 螺旋式压缩机
screw dislocation 螺位错，螺旋变位
screw feeder 螺旋进料机，螺旋加料器
screw mixer 螺旋式混合机
screw thread lubricant 螺纹润滑剂
screwy 扭曲的，螺旋形的
scribe 划痕［割］
scribing 划片，划割，划线
scrub 洗擦；用力擦洗，使净化
scrubbed 精制的，纯净的
scrubber 洗涤器，洗刷者，刷子
scrubbing 涤气，洗涤
scrubbing bottle 洗气瓶
scrubbing tower 涤气塔
scuff 磨损之处；使磨损
scuffing 变形，划痕，擦伤
sculpture 雕刻，刻蚀，风化
scum 浮渣，泡沫；产生泡沫，被浮渣覆盖，将浮渣去除掉
scumble 渐淡；将色彩弄淡，渐使颜色变浅，（薄涂）涂料
scummy 泡沫的，似（有）浮渣的，浮渣状的
scummer 消泡剂，除渣勺
seal 密封
seal welding 密封焊接

sealant 封闭剂
sealed tube 封闭管
sealed lead acid batteries 密封铅酸电池
sealing 封闭
sealing compound 嵌缝填料，密封剂，油灰，腻子
sealing failure 涂密封胶不良
sealing smut 封孔灰，挂灰
sealing test 封孔试验，密封试验
sealing wax 密封蜡
seam 缝；缝合，接合，裂开，产生裂缝
seaming 缝合，折弯重叠加工
seamless 无缝的，无缝合线的，无伤痕的
seamy 露出线缝的，有裂缝的
season cracking 季候缝裂，天然时效裂纹，老化开裂
seasoning 风干，自然干燥，气候处理，自然时效，陈化
sebaceous 分泌脂质的，脂肪的，脂肪质的，似油脂或皮脂的
sebacic acid 癸二酸，皮脂酸
second 秒，仲，第二的
second harmonic 二次谐波
second harmonic generation 二次谐波，倍频效应
second order phase change 二级相变
secondary 第二［中等，次要，中级］的
secondary accelerator 辅助促进剂
secondary alcohol 仲醇
secondary amine 仲胺，二级胺
secondary battery 次级［二次］电池
secondary carbon atom 仲碳原子
secondary charge effect 二次充电效应
secondary circuit 二次［次级］电路
secondary crystallization 二级晶化
secondary emulsifiers 次级乳化剂
secondary hardening 二次［回火］硬化
secondary ion mass spectrometry (SIMS) 二次离子质谱
secondary mechanical relaxations 二次机械松弛

secondary plasticizer 次级增塑剂，辅助塑化剂［增塑剂］
secondary reaction 副反应
secondary sedimentation tank 二次沉淀池
secondary treatment 二级［次级］处理
secondary valence 副价
section 型材
sectional 部分的，节的，可组合的
sectional titration 分区滴定
sectional view 断面［截面，剖面］图
sectrometer 真空管滴定计
sedentary 残积的
sedentary product 风化产物
sediment 沉淀物
sediment bowl 沉淀池［器］，澄清池
sedimental 沉积物的，沉降［冲积］的，由渣形成的
sedimentate 沉降，沉积
sedimentation 沉积，沉降积，淤积，沉降法
sedimentation analysis 沉淀分析
sedimentation coefficient 沉降系数
sedimentation constant 沉降常数
sedimentation equilibrium 沉降平衡
sedimentation potential 沉降电势
sedimentation velocity 沉降速度
sedimentation volume 沉降容积
sedimentator 沉淀器，离心器
seed crystal 晶种，籽晶
seep 漏，渗出［漏，滤］
seepage 渗透，渗漏，渗液
seepage loss 渗漏［渗透，渗出］损失
seepy 透水的，漏的
segregation 隔离，分离［隔，凝，层，开］，离解，偏析
seignette salt 罗谢尔盐，酒石酸钾钠
seizure 咬粘，滞塞
select appearance 选外观
selected ion monitoring 选择性离子检测
selective absorption 选择吸收
selective adsorbent 选择性吸附剂

selective corrosion	选择腐蚀，局部腐蚀
selective extraction	选择提取
selective hardening	部分淬火，局部淬火，选择硬化
selective oxidation	选择性氧化
selective reaction	选择反应
selective permeability	选择渗透性〔透过性〕
selective polymerization	选择聚合
selective reagent	选择试剂
selectivity	选择性，选择度
selectivity coefficient	选择系数
selectivity factor	选择系数
selenate	硒酸盐
selenate radical	硒酸根
selenic	硒的
selenic acid	硒酸
selenic anhydride	硒酐
selcnide	硒化物
selenious	亚硒的，二价硒的
selenite	亚硒酸盐
selenium	硒
selenium oxide	氧化硒
selenium rectifier	硒整流器
selenous acid	亚硒酸
selenylation	硒化
self cleaning	自净
self coagulation	自凝聚
self decomposition	自分解
self filtering	自滤
self passivation	自钝化
self quench hardening	自冷淬火
self tempering	自回火
self-assembly	自组装；自行组装的
self-catalyzed	自催化
self-colour anodizing	整体着色阳极氧化
self-combustible	自燃的
self-corrosion	自腐蚀
self-ignition	自燃
self-ignition temperature	自燃温度
self-induction	自感应，自诱导
self-lubricate	自（动）润滑
self-propagation hige-temperature synthesizing technology	自蔓延高温合成技术
self-purification	自净，自然净化
selfluminous pigment	自发光颜料
semi	半，部分，不完全
semi bright nickel	半光亮镍
semi dull	半光
semi-acetal	半缩醛
semi-automatic electroplating	半自动电镀
semi-automatic machine	半自动机器
semi-conductor detector	半导体探测器
semi-diaphanous	半透明的
semi-infinite condition	半无限条件
semi-integral	半悬挂式，半整体的
semi-mechanization	半机械化
semi-shearing	半剪
semi-solid processing	金属半固态加工
semi-translucent	半透明
semicarbazide	氨基脲
semiconductor	半导体
semiconductor rectifier	半导体整流器
semidestructive test	半破坏性测试
semidine	半联胺
semiebonite	半硬质胶
semifinished product	半成品，中间产品
semifolw sealing	半浮动涂密封胶
semigloss	半光泽的
semihydrate	半水合物
semimetal	半金属
semimicro	半微量的
semimicro analysis	半微量分析
semimicro balance	半微量天平
seminose	甘露糖
semipermeability	半透性，半透过性
semipermeable	半渗透的
semipermeable membrane	半透膜
semipolar	半极化
semipolar bond	半极性键
semipolar linkage	半极性键
semitransparency	半透明

sense-reversing 逆向的
sensibilisator 敏化剂
sensibility 敏感度，灵敏［精确，可感］度
sensibilization 敏化［增感］作用
sensibilizer 敏化剂
sensitive 敏感的，灵敏的，感光的
sensitive emulsion 感光乳剂［乳胶］
sensitive layer 感光层
sensitive paper 感光纸
sensitivity 灵敏度，敏感度，感光度
sensitivity control 灵敏度控制
sensitivity curve 灵敏度曲线
sensitivity limit 灵敏度限度
sensitivity speck 敏化中心
sensitivity value 灵敏度值
sensitization 敏化
sensitizer 敏化剂
sensitized fluorescence 增感发光，敏化荧光
sensitizing 光敏处理
sensitizing dye 敏化染料
sensitizing heat treatment 敏化热处理
sensitometer 感光计，曝光表
sensitometry 感光度测定
sensor 传感器
sensor array 阵列传感器
separalory funnel 分液漏斗
separate 分开；单独的，分开的；使分离［分开］
separated 分离出的，分开的
separating 分离的，分隔的
separating funnel 分液漏斗
separating power 分离能力
separation 分离
separation coefficient 分离系数
separation number 分离数
separative 分离性的，倾向分离的，区别性的
separatory 分离用的，使……分离的
sept- 【构词成分】七

septal 中隔［隔膜，间隔］的
septavalence 七价
septavalent 七价的
septivalent 七价的
sequence 次序
sequence number 序号，顺序数
sequence rule 次序规则
sequence test 顺序试验，序列试验
sequester 使隔绝［分离］
sequestrant 螯合剂，多价螯合剂作用，隐蔽剂
sequestration 隔离，隐蔽作用，价螯合剂作用
series connection 串联
serine 丝氨酸
serpentine 蛇状的，似蛇的，蛇形管
service column 运行塔，工作塔
sesqui- 【构词成分】一个半，一倍半
sesquioxide 倍半氧化物，三氧二某化合物
seston 浮游物，悬浮物
setting 装配，调整，凝结［固］，永久变形，收缩，硬化，下沉，沉降
setting accelerator 促凝剂，速凝剂
setting accuracy 定位精度，调整精度
setting agent 硬化剂
setting chamber 沉降室
setting point 凝固点
setting tank 固体沉降槽，硬化槽
setting time 凝结时间
settlement 沉降［积，渣］，澄清
settler 沉淀［沉降，滤清，澄清，分离，分级］器，沉降［积］槽
settling 沉淀［积，降］物，筛分，下沉
settling curve 沉降曲线
settling matter 沉淀物
settling ratio 沉降比
settling tank 沉降槽，沉淀池
settling vat 沉降桶，沉降槽
settling velocity 沉降速度
sewage 污水
sewage disposal 污水处理

sewage purification　污水净化，污水清化
sewage treatment　污水处理
sewage-disposal　污水处理
sewing　缝合，（塑料）熔合
sex-　【构词成分】六
sexmer　六聚物
sexangular　六角的，六边形的
sexavalence　六价
sextet　六重峰
shade　阴影处［部分］，（阴，黑）暗，遮光物，伞状物，（色彩）浓淡，明暗，深浅，色调，细微差别，不同程度
shading　底纹，遮蔽，明暗法
shaft furnace　井式炉
shakeout　落砂，出砂，筛选，离心分离
shakeout machine　落砂机
shallow cycle endurance　轻负荷寿命，轻负荷循环寿命
shallow pit defect　浅坑缺陷
shape defect　板型缺陷
shape memory alloy　形状记忆合金
shaping　成形加工；塑造的，成形的
sharp edge　利［锐］边
sharpener　磨具，削刀
sharp-set　批锋，使边缘锋利的
shaving　修面，削，修边
shear　切变，修剪；剪，剪切
shear crack　剪切裂缝，剪切裂纹
shear creep　剪切蠕变
shear elasticity　切变弹性，剪切弹性模量，切变模量
shear lip　剪切唇，切变裂痕
shear loading　剪切负荷，剪切载荷
shear modulus　剪切模量，剪切弹性模数
shear rate　剪切速率，切变速率
shear relaxation　剪切松弛，剪切弛豫
shear resistance　剪切阻力
shear strain　剪应变
shear strength　剪切强度，切变强度
shear stress　剪应力
shear waves　剪切波
shear thinning　剪切稀化
shear degradation　剪切退解
shearing action　剪切作用
shearing　剪，切断加工
shearing die　剪边模
shearing machines　剪切机
sheathing　防护物，外壳，入鞘
sheen　光辉［泽，彩］；光辉的，有光泽的，华丽的；闪耀，发光
sheeny　有光泽的，光亮的
sheet　薄板；片状的；成大片落下
sheet glass　板玻璃
sheet metal work　钣金工
shelf life　贮藏寿命
shell　外形，剥皮
shell and tube condenser　列管式冷凝器，管壳式冷凝器
shell and tube heat exchanger　列管式换热器
shell casting　壳模铸造
sherardizing　粉镀锌法
sherwood oil　石油醚
shield　防护物，屏蔽；遮蔽；防御，起保护作用
shield gas　保护气体
shield jig　保护夹具
shielding constant　屏蔽常数
shielding effect　屏蔽效应，屏蔽作用
shielding factor　屏蔽因子［因数，系数］
shielding solvent　掩蔽溶剂
shift　移动，偏移，转移
shim　薄垫片
shim plate　垫板
shining nickel　光亮镍
shiny　有光泽的，擦亮［闪耀］的，磨损［光］的
ship bottom paint　船底涂料，船底漆
shoal　变浅；使变浅；浅的
shock line　模口挤痕
shock load　冲击荷载
shock resistance　耐冲击性，耐震强度

shockproof 防震的，防电击的
Shore hardness tester 肖氏硬度计
short 短路；短［不足，矮，低］的；不足
short-brittle 热脆
short circuit checking 测短路
short-circuit current 短路电流
short fiber 短纤维
short oil varnish 短油清漆，稀油漆
short shot 充填不足
short-time discharge 短时间放电
short-time fourier transformation (STFT) 短时傅里叶变换
shot blast 喷丸处理，喷砂清理
shot blast chamber 喷丸室
shot blasting 喷丸处理，珠粒喷击清理
shot peening 喷丸加工，珠击处理
shoulder peak 肩峰
shredding （机械）裂解，粉碎（作用），研末（作用），纤化
shrend 水淬
shrinkage 收缩，减低
shrinkage cavity 缩孔
shrinkage hole 缩孔，收缩孔
shrinkage limit 收缩限度［极限］
shrinkage pool 凹孔
shrinkage volumetric 容量收缩
shrinkage water 收缩水
shrinkage crack 缩裂，收缩裂缝
shrinking percentage 收缩率，收缩百分比
shrinking stress 收缩应力
shrinkproofing 防缩
shroud 掩蔽，屏板，遮盖，隐蔽，遮蔽；覆盖
shunt 分流器；使分流
siccative 干燥剂；使干燥的
side chain isomerism 侧链异构
side chain 侧链
side elevation 侧视图
side etching 侧面蚀刻，侧蚀
side reaction 副反应
side reaction coefficient 副反应系数
side reciprocator 侧喷机
side scrap 切边
side stream filtration 旁流过滤
side stretch 侧冲压平
side view 侧视图
side wall 侧面［壁］
siderophile element 亲铁元素
sieve 筛子，滤网；筛，滤
sieve analysis 筛析，筛分分析
sieve mesh 筛孔，筛眼
sieving machine 筛选［分］机
sieving 筛分［选］
signal generator 信号发生器
signal gradient 信号梯度
signal overload point 信号过载点
signal shot noise 散粒噪声
signal strength 信号强度
signal to noise ratio 信噪比，信号噪声比
significant test 显著性检验
silaceous 含硅的
silane 硅烷，矽烷
silane coupling 硅烷耦合
silanol 硅烷醇，硅醇
silic- 【构词成分】硅
silica 二氧化硅，石英
silica gel 硅胶
silica glass 石英玻璃
silica sand 石英砂
silicasol 硅溶胶
silicate 硅酸盐
silicate paint 硅胶漆
silicate structure 硅酸盐结构
silicatization 硅化，硅化作用
siliceous 硅酸的，硅土的
silicic 硅（酸，石）的
silicic acid 硅酸
silicide 硅化物
silicious 含硅［硅土，硅酸］的
silico- 【构词成分】硅
silicofluoride 氟硅化物，氟硅酸盐

silicon 硅
silicon acid 硅酸
silicon carbide 碳化硅
silicon carbide fiber 碳化硅纤维
silicon carbide paper 水磨砂纸
silicon chloride 氯化硅
silicon dioxide 二氧化硅
silicon fluoride 氟化硅
silicon hydride 硅氢化合物，硅烷
silicon nitride 氮化硅
silicone 硅酮，硅树脂
silicone oil 硅酮油，硅油
silicone paint 硅树脂漆
silicone resin 有机硅树脂
silk screen 丝印
silk screening on metal parts 金属产品丝印
silk-screen printing 丝网印刷
silk-screen process 丝网印刷法，绢印法
silken 绸的，柔软的，丝制的
silkiness 柔软，绸缎般
siloxane 硅氧烷
silver 银
silver acetate 醋酸银
silver acetylide 乙炔银
silver amalgam 银汞剂
silver arsenite 亚砷酸银
silver bromide 溴化银
silver brushed finish 银拉丝，磨砂面纯银效果
silver cadmium batteries 银钙电池
silver carbide 碳化银
silver carbonate 碳酸银
silver chloride 氯化银
silver chloride cell 氯化银电池
silver chromate 铬酸银
silver compound 银化合物
silver cyanide 氰化银
silver fluoride 氟化银
silver foil 银箔
silver halide 卤化银

silver ion 银离子
silver nitrate 硝酸银
silver oxide 氧化银
silver oxide batteries 银氧化物电池
silver plating 镀银
silver salt 银盐
silver-sliver chloride electrode 银-氯化银电极
silvered 镀银的
silvering 镀银，所镀的银层，银色光泽
silvern 银一般的，银的
similarity 类似性，相似点
similitude 相似，外表，比拟
simulated 模拟的，模仿的，仿造的
simulated annealing 模拟退火
simulation 模拟［仿］，仿真
simultaneous 同步的，同时发生的
single bond 单键
single crystal 单晶
single crystal making 单晶制备
single electrode 单电极
single electron transfer 单电子转移
single fluid cell 单液电池
single phase system 单相系
single side lapping machine 单面磨光机
single side polishing machine 单面抛光机
single stage nitriding 等温渗氮
single-sweep polarography 单扫描极谱法
singlet 单峰
sink 水槽．洗涤槽．污水坑；下沉，渗透
sinkhole 污水坑，排水口，阴沟口
sinking 沉没，凹陷，下沉的
sinter 烧结；使烧结［熔结］
sinter forging 烧结锻造
sintering 烧结处理；烧结的
sintering furnace 烧结炉
sinuate 波状的，弯弯曲曲的；弯曲
sinuosity 弯曲，弯曲处，曲折度
sinusoid 正弦曲线
sinusoidal 正弦曲线的
sinusoidal waveform 正弦波形

siphon	虹吸管；用虹吸管吸出，抽取
siphonage	虹吸作用，虹吸能力
siphonal	虹吸[水管]的，虹吸作用的
six membered ring	六元环
six sides forging	六面锻造
size analysis	粒度分析
sizing	胶[涂]料，涂上胶水
sizing agent	上浆剂，施胶剂
sizy	胶水[糨糊，黏稠]的
skeller	挠曲（变形）
sketch	草图
skew	斜交，歪斜；斜交[歪斜]的
skewness	歪斜，不对称度，偏度
skim	撇，撇去的东西，表层物；脱脂的，撇去浮沫的，表层的；撇去……的浮物
skim milk	脱脂乳
skimmer	撇渣[沫，油，分液]器，挡渣芯
skin friction	表面摩擦
skin inclusion	表皮折叠
skinning	油漆起结皮
skip	漏，遗漏
skive	把……割成薄片，削匀，研磨
skiving	刮削，表面研磨，切片
slabby	板状的，黏稠的
slack	松弛的；放松
slack lime	熟石灰，消石灰
slack quenching	断续淬火
slag	炉[渣，熔]渣；使成渣，使变成熔渣
slag action	炉渣侵蚀，熔渣侵蚀作用
slagging medium	助熔剂
slaking	水[消，熟]化，潮解
slaking value	水化[消化]值
slant	倾斜；使倾斜；倾斜的
slash	长缝，螺纹深压
sleak	冲淡，溶化
sleeve	套筒；给……装套筒
slice	薄片，部分；切下，把……分成部分，将……切成薄片
slicing machine	切割[切片]机
slickenside	（断面）擦痕
slidabrading	滚光
slide	滑动；使滑动
sliding friction	滑动摩擦
slime	泥渣；涂泥
slip	滑动的，有活结的；使滑动，滑过；
slip agent	助滑剂
slip plane	滑动[滑移]面
slipper dipping	拖式浸涂
slit	缝隙，裂缝；使有裂缝
slitless	无缝的
sliver	裂片；使成薄片，使裂成小片
slot	[裂，狭]槽，槽沟
slotted	有沟槽的，有裂痕的
slotting	切缝[削]，开槽
slow oxidation	缓慢氧化
slow strain rate	慢应变速率
slow vent	缓慢通气
slowing down	减速，慢化
sludge	泥渣，污泥
sludge digestion	污泥消化
sludge incineration	污泥焚烧
sludge integrated application	污泥综合利用
slug feed	批量投药
sluice	水闸，蓄水，洗矿槽；冲洗，开闸放水
slurry	泥[稀，砂，釉]浆，悬浮液
slush molding	凝塑成形
slushing oil	防锈油，抗蚀油
smelt	熔融[炼]，冶[精]炼
smelting	冶炼
smelting furnace	熔炉，冶炼炉
smirch	沾污，污迹；弄脏
smoke agent	烟雾剂
smooth surface	光面
smoothness	平滑度，柔滑，平坦
smudge	污点[迹]；弄脏，涂污
soak	浸，湿透；浸泡[透]，吸收
smudginess	污物
soakage	浸（渍），浸湿性

soaking 均热处理，浸热
soap 皂
soap builder 肥皂配合剂
soap powder 皂粉
soap-treated 用肥皂处理过的
soda 苏打
soda ash 苏打灰
soda feldspar 钠长石
soda lime 碱石灰
soda soap 钠皂
sodalye 氢氧化钠
sodden 浸透的；使浸透
sodion 钠离子
sodium 钠
sodium acetate 醋酸[乙酸]钠
sodium alkyl benzene sulfonate 烷基苯磺酸钠
sodium alkyl sulfate 烷基硫酸钠
sodium alkylsulfonate 烷基磺酸钠
sodium aluminate 铝酸钠
sodium arsenate 砷酸钠
sodium arsenite 亚砷酸钠
sodium bicarbonate 碳酸氢钠，小苏打
sodium bichromate 重铬酸钠，红矾钠
sodium bisulfate 硫酸氢钠
sodium bisulfite 亚硫酸氢钠
sodium borohydride 硼氢化钠
sodium bromate 溴酸钠
sodium bromide 溴化钠
sodium carbonate 碳酸钠
sodium carboxymethyl cellulose (CMC) 羧甲基纤维素钠
sodium cellulose glycolate 乙酸钠纤维素
sodium chlorate 氯酸钠
sodium chloride 氯化钠
sodium chlorite 亚氯酸钠
sodium chlorite bleaching 亚氯酸钠漂白
sodium chromate 铬酸钠
sodium citrate 枸橼酸钠，柠檬酸三钠，2-羟基丙烷-1,2,3-三羧酸钠二水合物
sodium cyanide 氰化钠
sodium dichromate 重铬酸钠
sodium dihydrogenphosphate 磷酸二氢钠
sodium dithionite 亚硫酸氢钠，连二亚硫酸钠
sodium dodecyl benzene sulphonate 十二烷基苯磺酸钠
sodium dodecyl sulfate (SDS) 十二烷基硫酸钠
sodium ethoxide 乙醇钠
sodium ferricyanide 铁氰化钠
sodium fluoride 氟化钠
sodium formate 甲酸钠
sodium hexametaphosphate 六偏磷酸钠
sodium hydrogencarbonate 碳酸氢钠
sodium hydrogensulfate 硫酸氢钠
sodium hydrosulfite 亚硫酸氢钠
sodium hydroxide 氢氧化钠
sodium hypochlorite 次氯酸钠
sodium hypophosphite 次磷酸钠
sodium lauryl sulfate 十二醇硫酸钠，月桂醇硫酸酯钠盐
sodium lauryl sulphate 十二烷基硫酸钠，月桂基硫酸钠
sodium metaphosphate 偏磷酸钠
sodium metasilicate 偏硅酸钠
sodium molybdate 钼酸钠
sodium nitrate 硝酸钠
sodium nitrite 亚硝酸钠
sodium oxide 氧化钠
sodium perborate 过硼酸钠
sodium peroxide 过氧化钠
sodium phosphite 亚磷酸钠
sodium polysulfide 多硫化钠
sodium potassium tartrate 酒石酸钾钠
sodium pyrophosphate 焦磷酸钠
sodium pyrosulfite 焦亚硫酸钠
sodium salicylate 水杨酸钠
sodium selenite 亚硒酸钠
sodium silicate 硅酸钠
sodium silicofluoride 氟硅酸钠
sodium stearate 硬脂酸钠

sodium sulfate	硫酸钠
sodium sulfide	硫化钠
sodium sulfite	亚硫化酸钠
sodium tetraborate	四硼酸钠
sodium thiocyanate	硫氰酸钠
sodium thiosulfate	硫代硫酸钠
sodium tripolyphosphate (STPP)	三聚磷酸钠
sodium tungstate	钨酸钠
soft magnet material	软磁材料
soft potassium soap	软钾皂
soft water	软水
soft-drawn	微拉伸，软拔[拉，抽]的
softener	软化剂，增塑剂，软化装置
softening	软化，变软
softening agent	软化剂
softening degree	软化度
softening point	软化点
softening temperature	软化温度
soil	土壤；弄脏，变脏，污染
soil corrosion	土壤腐蚀
soil stabilizer	土壤稳定剂
sol	溶胶
solar distillation	曝晒蒸发
solar drying	日晒干燥
solar radiation	太阳[日光]辐射
solar still	太阳能蒸馏[蒸发]器
solarization	老化作用，日晒
solation	溶胶化
solder	焊剂；焊接，使连接在一起
solderability	可焊性，钎焊性
solderability test	焊接性测试，可焊性试验
soldering	焊接[料，接处]；用于焊接的
soldering acid	钎焊用酸，钎焊液
soldering pinhole	焊药针孔
solid color	本色
solid dilution	固溶体
solid electrolyte	固体电解质
solid fuel	固体燃料
solid liquid equilibrium	固液平衡
solid lubricant	固体润滑剂
solid matter	固体物质
solid paraffin	固态石蜡
solid phase	固相
solid phase polymerization	固相聚合
solid phase reaction	固相反应
solid solution	固溶体
solid solution strengthening	固溶强化
solid state	固态
solid state plasma	固态等离子体
solidification	凝固，浓缩，变浓，结晶
solidified	凝固的，固化的，变硬的
solidifying point	凝固点
solidoid	固相
solidus	固相线
soliquoid	悬浮体
solubility	溶解度
solubility curve	溶解度曲线
solubility limit	溶解限度
solubility parameter	溶解度参数
solubility product	溶度积
solubility product constant	溶度积常数
solubilization	加[增]溶，溶化
solubilize	溶液化，增溶（化），溶解
solubilizer	增溶剂
solubilizing ability	溶化能力
solubilizing power	增溶力
soluble	可溶的
soluble saccharin	可溶性糖精
solute	溶质
solute-solvent interaction	溶质-溶剂相互作用
solution	溶液，固溶处理
solution coating	溶液涂膜
solution control	槽液管理
solution condensation	溶液缩合
solution copolymerization	溶液共聚合
solution dyeing	溶液染色
solution extrusion	溶液挤压
solution measurement	溶液测定
solution polymerization	溶液聚合

solution pressure	溶解压力
solution property	溶液特性
solution technique	溶液技术
solution treatment	固溶处理，固溶化热处理
solution viscosity	溶液黏度
solutrope	共溶混合物，相溶物
solvable	溶解的，可溶的
solvate	溶剂化物
solvated electron	溶剂化电子
solvation	溶解，熔化，溶剂化作用
solvent	溶剂
solvent cleaner	溶剂清洗剂
solvent degreasing	有机溶剂除油，溶剂脱脂
solvent effect	溶剂效应
solvent extraction	溶剂浸出［提取，萃取］
solvent molding	溶剂成型
solvent parameter	溶剂参数
solvent recovery	溶剂回收
solvent refining	溶剂精制
solvent remover	溶剂去除剂，溶剂洗净剂
solvent resistance	耐溶剂性
solvent selection	溶剂选择
solventless coating	无溶剂型涂料，无溶剂涂膜
solventless varnish	无溶剂清漆
solvent-extracted	溶剂萃取的
solvent-hating	疏液的
solvent-removal penetrant	溶剂去除型渗透剂
solvolysis	溶剂分解
somorphous	匀晶相图
sonim	夹砂
sorb	吸附，吸收
sorbate	吸着物，山梨酸酯
sorbefacient	吸收促进剂；促进吸收的
sorbent	吸附剂
sorbic acid	山梨酸，己二烯酸
sorbierite	山梨醇
sorbin	山梨糖
sorbitan	山梨聚糖，脱水山梨糖醇
sorbitol	山梨糖醇
sorbose	山梨糖
sorption	吸着，吸附作用
sorption capacity	吸附能力
sorption equilibrium	吸着平衡
sorption film	吸附膜
sorption-extraction	吸附［离子交换］提取
sorrel	栗色；栗色的，红褐色的
sosoloid	固溶体，固态溶液
source	源
space isomerism	空间异构
space lattice	空间晶格，空间点阵
space model	空间模型，立体模型
space polymer	立体聚合物
spacer	隔板
spacing	间距
spacing washer	间隔垫圈
spanner	扳手，螺丝扳手
spar varnish	桅杆清漆，（船舶）底漆
spark	火花；闪烁，发火花
spark discharge	火花放电
sparkle silver	闪银
specked	有斑点的
speckless	无斑点的，无瑕疵的
speckling	毛刺，麻点；弄上斑点，点缀
species	种类，物质；物种上的
specific abrasion energy	比磨耗能
specific activity	比活性
specific adsorption	特性吸附，比吸附
specific charge	比电荷
specific cohesion	比凝聚力
specific conductance	比电导
specific conductivity	比电导率
specific gravity	相对密度
specific gravity balance	比重天平
specific gravity bottle	比重计
specific gravity cup	比重杯
specific gravity indicator	比重计

specific ionization	比电离
specific resistance	比电阻
specific strength	相对强度
specific surface area	比表面积
specific weight	相对密度
specification	规范，规格，质量指标
specificity	特异性
specified achromatic light	标准光源
specified sensitivity	给定灵敏度，给定感光度
specimen	样品，样本，标本
spectral	光谱的
spectral analysis	光谱分析
spectral characteristics	光谱特性
spectral distribution	光谱分布
spectral line	光谱线
spectral sensitivity	光谱感度
spectral series	光谱系
spectrochemical analysis	光谱化学分析
spectrochemistry	光谱化学
spectroelectrochemistry	光谱电化学，分光电化学
spectrograph	摄谱仪
spectrometer	光谱计，分光仪
spectrometer dispersion	谱仪色散
spectrometric	光谱测定的，度谱的，分光仪的，光谱仪的
spectrometry mass	质量分析仪
spectrophotometer	分光光度计，分光计
spectrophotometry	分光光度法
spectroscope	分光镜
spectroscopic analysis	光谱分析
spectroscopic	光谱学的，分光镜的
spectrum	光谱
spectrum analysis	光谱分析
spectrum band	光谱带
specular	镜子的，会反射的，镜子一般的
specular gloss	镜面光泽度
specular reflection	镜面反射，镜面光泽度
spent acid	废酸
spent lye	废碱液
spermaceti	鲸蜡
spherical	球形的，球面的
spherical condenser	球形冷凝器
spherical sector analyser (SSA)	球扇型能量分析器
spheroidal	类似球体的，球状的
spheroidicity	球形，扁球体状，椭球体状
spheroidite	球状渗碳体，粒状化
spheroidization	球状化处理
spheroidize	球化处理，延期热处理
spheroidizing annealing	球化退火
spherulite	球晶
spilliness	（钢丝表面缺陷）鳞片，毛刺
spilth	溢出，溢出物，剩余物
spin	自旋
spin dryer	旋转式脱水机
spin finishing	旋转研磨
spindle	轴，锭子；锭子的，锭子似的；装锭子于
spindle oil	锭子油
spinel	尖晶石
spiral	螺旋，螺旋形之物；螺旋形的；使成螺旋形，使作螺旋形上升
spiral condenser	旋管冷凝器
spiral contractometer	螺旋应力仪，螺旋收缩力计
spiral flow test	螺旋流动试验
spirit color	醇溶染料
spirit varnish	醇溶清漆
spit-out	喷孔
spitting spray	气喘，喷流间断
splash	飞溅的水，点；溅，泼，用……使液体飞溅
splash guard	防溅罩，防溅板
splay	展开，斜面；倾斜的，八字形的
spline	花键，齿条，塞缝片；开键槽，用花键连接
split	劈开，裂缝；劈开的；分离，劈开，离开，分解
split product	裂解产物
split regeneration	分流再生

sponge 海绵
sponginess 海绵质，海绵状
spontaneous 自发的，自然的
spontaneous crystallization 自发结晶
spontaneous magnetization 自发磁化
spontaneous processes 自发过程
spot analysis 点滴分析
spot annealing 局部退火
spot corrosion 点腐蚀，点蚀
spot reaction 点滴反应
spot repair 局部修补
spot test 抽查，当场测试
spot welding 点焊
spotted 有斑点的，斑纹的，弄污的
spout 流出口；喷出［射］
spray 喷雾，喷雾器，水沫；喷射
spray booth 喷漆室
spray coating 喷涂
spray dryer 喷雾干燥器
spray dyeing 喷雾染色
spray gum tip 涂料喷嘴
spray gun 喷枪
spray gun velocity 喷枪移动速度
spray lacquer 喷漆
spray mist compatibiliity 抗漆雾污染性
spray nozzle 喷嘴
spray polymerization 喷雾聚合
spray pretreatment system 喷射式前处理
spray rinsing 喷射清洗
spray sizing 喷雾上浆
spray system pretreatment equipment 喷射式前处理装置
spray tower 喷雾塔
spray water proofing 喷雾防水
spraying 喷涂，喷雾，喷敷法
spraying jet 喷射流
spread 传播，伸展；伸展的；传播，散布
spreading agent 铺展剂
spreading ability 铺展能力
spreading rate 涂布率，扩张速率
spring 弹簧

spring balance 弹簧秤
spring constant 弹簧系数［常数］
spring steel 弹簧钢
spume 泡沫；喷吐（泡沫），起泡沫
spumescence 泡沫状，起泡沫
spumescent 起泡沫的，泡沫状的
spur 毛刺
spurging 起泡，产生泡沫
sputter 溅射，阴极针孔喷镀，阴极溅镀
sputter coating 溅射镀膜
sputter depth profile (SDP) 溅射深度剖析
sputter etching 溅射蚀刻
sputter-gun sputtering system 溅射枪溅镀系统
sputtering 溅射，喷溅涂覆法，阴极真空喷镀
sputtering rate 溅镀速率
sputtering system 溅镀系统
sputtering yield 溅射效率
square wave 方波
square wave polarography 方波极谱法
square wave voltammetry 方波伏安法
squeezable 可压缩的
squeeze casting 模压铸造
stability 稳定性
stability constant 稳定常数
stability test 稳定性试验
stabilization 稳定化
stabilizer 稳定剂
stabilizing treatment 稳定化热处理
stacking fault 叠层缺陷，堆垛层错
stactometer 滴量计
stagger 交错安排；交错的，错开的；使交错
staggered 错列［交错，叉排，参差］的
stain 着色剂，玷污
stain resistance 涂膜污染性，耐污染性
stainable 可染色的
stained 玷污的，染色的，有斑点［纹］的

staining 着[染]色，污染，锈蚀，侵蚀	stannate 锡酸盐
stainless steel 不锈钢	stannic 锡的，四价锡的，含锡的
standard electrode potential 标准电极电位	stannic acid 锡酸
staircase voltammetry 阶梯伏安法	stannic chloride 氯化锡
stalagmometer 表面张力（滴重计）	stannic oxide 二氧化锡
stamp letter 冲字（料号）	stannic salt 正锡盐
stamp mark 冲记号	stanniferous 含锡的
stamping 冲压	stannize 渗锡
stamping part 冲压件	stannous 锡的，含锡的，含二价锡的
stamping tool 冲压工具	stannous chloride 二氯化锡
stanch 防水[密封，气密]的；使停止流溢	stannous hydroxide 氢氧化亚锡
standard 标准	stannous oxide 氧化亚锡
standard addition 标准加入法	stannous salt 亚锡盐
standard calomel electrode 饱和甘汞电极，标准甘汞电极	starch 淀粉；给……上浆
standard cell 标准电池	starch paper 淀粉试纸
standard curve 标准曲线	starch-iodine reaction 淀粉碘反应
standard deviation 标准偏差	static 静态的，静电的，静力的
standard electrode potential 标准电极电位	static bed 固定床
standard electromotive force 标准电动势	static charge 静电荷
standard equilibrium constant 标准平衡常数	static electricity 静电
standard hydrogen electrode 标准氢电极	static electrode potential 静态电极电位
standard illuminant 标准光源	static friction 静摩擦
standard part 标准件	static mercury drop electrode (SMDE) 静态汞滴电极
standard potential 标准电势	static pressure 静压
standard pressure 标准压力	stationary 固定的，静止的
standard rate constant 标准速率常数	stationary mercury electrode 固定汞电极
standard sample 标准试样	stationary phase 固定相
standard sand 标准砂	steady state diffusion 稳态扩散
standard solution 标准溶液	steam 水蒸气
standard solution 标准溶液	steam bath 蒸汽浴
standard state 标准状态	steam boiler 蒸汽锅炉
standard temperature 标准温度	steam calorimeter 蒸汽热量计
standard test block 标准试块	steam coil 蒸汽旋管，蛇形蒸汽管
standard test method 标准试验方法	steam condensate 蒸汽冷凝水
standardization 标定	steam consumption 蒸汽消耗，耗汽量
standardization instrument 标准化仪器	steam distillation 蒸汽蒸馏
standard deviation 标准偏差	steam drier 蒸汽干燥器
stann- 【构词成分】锡	steam ejector 蒸汽喷射器
	steam emulsion number 蒸汽乳化值
	steam funnel 蒸汽漏斗

steam generator 蒸汽发生器
steam heating 蒸汽加热
steam jacket 蒸汽套
steam jet 蒸汽喷射
steam phosphating 蒸气磷化处理
steam sealing 水蒸气封孔处理
steam sterilizer 蒸汽杀菌器
steam trap 汽水阀
steam valve 疏水阀，疏水器，汽水分离器，凝汽阀
steam vulcanization 蒸汽硫化
steaming 蒸汽加工；冒热气的
stearate 硬脂酸盐
stearic acid 硬脂酸，十八酸
stearin 硬脂，硬脂酸甘油，硬脂酸甘油酯
stearin pitch 硬脂沥青
stearyl 硬脂酰，十八烷酰
stearyl alcohol 硬脂醇，十八烷醇
stechiometer 化学计量仪
steel 不锈钢
steel ingot 钢锭
steel plate 钢板
steel spatula 钢刮刀
steel wire 钢丝
steep 浸渍；急剧升降的；泡，浸
steeping 浸渍的
stepwise 逐步地，阶梯式地
stepwise regeneration 分步再生
stereo isomer 立体异构体
stereochemical 立体化学的
stereochemical formula 立体化学式
stereochemical orientation 立体（化学）取向
stereoselective 立体选择的
stereoselective reaction 立体选择反应
stereoselectivity 立体选择性
stereospecific synthesis 立体定向合成
stereospecificity 立体特异性
stereostructure 立体结构
steric 位的，（原子的）空间（排列）的

steric effect 位阻效应
steric exclusion chromatography 空间排阻色谱法
steric factor 位阻因素
steric hindrance 位阻，位阻现象
steric strain 空间应变，立体应变
stib- 【构词成分】锑
stibate 锑酸盐
stibial （正，五价）锑的，像（正）锑的
stibiate 锑酸盐
stibiated 含锑的
stibic 锑的
stibide 锑化物
stibious 含三价锑的
stick tape 贴胶带
stickiness 黏滞性，胶黏
sticking agent 黏着剂
stiffener 加强剂
stir 搅动
sticking 黏［滞，吸］附作用，黏（结，胶，料，模）
stirrer 搅拌机
stirring 搅拌
stirring equipment 搅拌装置
stochastic 随机
stochastic variable 随机变量
stock 毛坯，坯料
stock solution 储液
stockpile 贮存，库存
stoichiometric 化学计量的，化学计算的
stoichiometric calculation 化学计算
stoichiometric point 化学计量点
stoichiometrically 化学计算（计量，当量，数量）地
stoichiometry 化学计量学
storage 储藏
storage loss 储存损失，仓储损失
storage stability 储藏稳定性，耐储存性
straight chain 直链
straight run distillation 直接蒸馏
straight run gasoline 直馏汽油

straightening	矫直
straightening annealing	矫直退火
straightening machine	矫直机
strain	应变，变形，延伸率；拉紧，拉长，压缩（变形），使变形
strain ageing	形变时效
strain energy	应变能
strain strengthening	形变强化
strainer	滤网，过滤器
strait-	【构词成分】层
stratification	层理，成层
stratification sampling	分层取样
stratified	分层的，有层次的，层状的
stratiform	层状的，成层的
stratify	分层，成层
strap	带；用带捆绑
stratographic	色层分离的，色谱的
stratography	色层分离，色谱法
stray	杂散的，分散的
stray current	杂散电流
stray current corrosion	杂散电流腐蚀
stray light	杂散光
streak	条状痕，条纹
streaked	有条纹的；在……上加条纹
streaming potential	泳动电势，流动电位
streamline flow	层流，流线流，流线型流动
strength	强度
strength test	强度试验
stress	应力
stress concentration	应力集中
stress corrosion	应力腐蚀
stress corrosion cracking	应力腐蚀断裂
stress cracking	应力破裂
stress decay	应力衰变
stress intensity factor	应力场强度因子
stress relaxation	应力松弛
stress relief	消除应力，应力释放
stress relieving annealing	应力消除退火
stress strain diagram	应力，应变图
stress-induced phase transformation toughening	应力诱导相变增韧
stretch	拉伸，延伸；伸展，张开，可伸缩的
stretching test	拉伸试验
stretchability	可延性，拉伸性
stretching vibration	伸缩振动，弹性振动
striation	起条纹，生长条纹
strike plating	冲击镀
strike	预镀
striking current	冲击电流
stripe	条纹，斑纹；加条纹于……
strippable	可剥［移，拆，取］去的
strippable coating	可剥性涂料
strippant	洗涤剂，解吸剂
stripper	（冲孔）模板，脱模机，分离装置，剥皮器，涂层消除器
stripping	破裂，拆开，除去，剥皮［离，开，落，裂］，去皮［膜，胶，色］，冲掉，退镀
stripping agent	退色剂，剥离剂
stripping analysis	溶出分析
stripping film	可剥膜
stripping method	溶出法
stripy	有条纹的，条纹状的
strong acid	强酸
strong base	强碱
strong electrolyte	强电解质
strontium	锶
strontium carbonate	碳酸锶
struction	结构式
structural	结构的
structural alloy steel	合金结构钢
structural attachment	结构附件
structural isomerism	结构异构
structural phase transition	结构相变
structure	构架，结构
structure analysis	结构分析
stuffing	填料，填塞物
stuffy	不通气［风］的
stylo-	【构词成分】尖（头）的
styrene	苯乙烯

styrone	肉桂醇,苯乙烯树脂
subacid	稍带酸味的,微酸性的
subbottom	底基
sub-boundary	亚晶界,小晶粒间界,粒界网状组织
subcarbonate	碱式碳酸盐
subcell	亚晶胞
suberate	辛二酸盐
suberic acid	辛二酸
sublattice	子格,亚晶格
sublimability	升华性,升华能力
sublimable	可升华的
sublimate	升华物;使升华,纯化;纯净化的
sublimation	升华
sublimation cooling	升华冷却
sublimation curve	升华曲线
sublimation point	升华点
sublime	使……纯化,升华,纯化
submergence	下沉,淹[浸,沉]没
submicro	亚微米
submicro analysis	半微分析
submicro metal	超微[细]金属粉末
subsidiary	附属的,辅助的
subsidiary valence	副价
subsiding tank	沉淀槽
subsoil water	地下水
substance	物质
substantive color	直接染料
substituent	取代基
substitution	取代
substitutional	取代的,代用的
substitutional type solid solution	置换型固熔体
substractively normalized interfacial Fourier transform infrared spectroscopy (SNIFTIRS)	差减归一化界面傅里叶变换红外光谱学
substrate	底材,基底
substrate mental	基体金属
substructure	亚组织
subsurface	地下的,表面下的
subsurface corrosion	皮下腐蚀
subsurface discontinuity	近表面不连续性,表面下缺陷
subtract	减,减去,扣除
subzero treatment	生冷处理
successive reaction	逐次反应
succinaldehyde	琥珀醛,丁二醛
succinamide	琥珀酰胺,丁二酰胺
succinate	琥珀酸盐,琥珀酸,丁二酸
succinic acid	琥珀酸,丁二酸
sucrose	蔗糖
suction	吸收
suction filter	吸滤器
suction pump	抽吸泵
sufficient	足够的,充分的
sulfa-	【构词成分】磺胺
sulfamic acid	氨基磺酸
sulfamide	硫酰胺,磺酰胺
sulfanilamide	对氨基苯磺酰胺,磺胺
sulfanilic acid	磺胺酸
sulfate	硫酸盐;使成硫酸盐,用硫酸处理
sulfate oil	磺化油
sulfate process	硫酸盐法
sulfathiazole	磺胺噻唑
sulfation	硫酸化,硫酸盐化作用
sulfenylation	亚磺酰化
sulfidation	硫化
sulfide	硫化物
sulfinic acid	亚磺酸
sulfitation	亚硫酸盐化
sulfite	亚硫酸盐
sulfite process	亚硫酸盐法
sulfite waste liquor	亚硫酸盐废液
sulfobenzoic acid	磺基苯酸
sulfofication	硫化作用
sulfogroup	磺基
sulfonamide	磺酰胺
sulfonate	磺酸盐
sulfonation	磺化(作用)
sulfonation reaction	磺化反应

sulfone 砜
sulfonic acid 磺酸
sulfonic acid group 磺（酸）基
sulfonylation 磺酰化
sulfosalicylic acid 磺基水杨酸
sulfoxide 亚砜
sulfoxylate 次硫酸盐
sulfoxylic acid 次硫酸
sulfur 硫，硫黄
sulfur bacteria 硫细菌
sulfur dioxide 二氧化硫
sulfuration 硫化
sulfuric acid 硫酸
sulfuric acid anhydride 硫酐
sulfurous acid 亚硫酸
sulfurous anhydride 亚硫酐
sulfurous ester 亚硫酯
sulfuryl chloride 硫酰氯
sulphate radical 硫酸根
sulphate 硫酸盐
sulphation 硫酸化
sulphide 硫化（物）；变成硫化物，用硫化物处理
sulphiteaddition 亚硫酸盐加成
sulphion 硫离子
sulphitation 亚硫酸化（作用）
sulphite 亚硫酸盐；用亚硫酸（盐）处理
sulpho- ［构词成分］硫（代），磺基
sulphoacid 磺酸
sulphoacid anodic oxide coating 硫酸阳极氧化膜
sulphonation 磺化
sulphur 硫黄；使硫化，用硫黄处理
sulphur dioxide test 二氧化硫试验
sulphuric acid 硫酸
sulphurize 用硫处理，使硫化
sulphurizing 渗硫
sump 水坑，污水坑，机油箱
supercarbonate 碳酸氢盐
supercarburize 过度渗碳
supercavitation 超空化

superchlorination 过氯化，过氯化作用
supercompressibility 超压缩性
superconducting material 超导材料
superconductivity 超导性
supercontraction 超缩，超收缩性
supercooled liquid 过冷液体
supercooling 过冷
supercritical fluid chromatography (SFC) 超临界流体色谱法
superfinish 超级光泽，超精表面，超精加工
superheated steam 过热蒸汽
superimpose 叠加，添加，信息叠加
superimposed current electroplating 叠加电流电镀
superimposed load 超载
superimposition 叠印［加］，重叠
superindividual 超单晶
superinsulant 超绝缘体
supermicro analysis 超微分析
supermolecule 超分子，胶束，微胞
superoxide 过氧化物，超氧化物
superoxol 过氧化氢溶液（30％）
superpolymer 高聚物
super-resolution 超分辨率
supersaturate 使过度饱和
supersaturated solution 过饱和溶液
supersaturated steam 过饱和蒸气
supersaturation 过饱和
supersonic 超声波的
supersonic wave 超声波
supersymmetry 超对称性
supplementary 补充的，追加的
supplementary requirements 附加要求
supply duct 供气风管
supplying fan 供气风机
support point 支持点
supported liquid membrane (SLM) 支撑液膜
supporting electrolyte 支持电解质
suppression 抑制

surcharge 超载；使……装载过多
surface 表面
surface plasmon resonance (SPR) 表面等离子体共振
surface-enhanced infrared adsorption (SEIRA) 表面增强红外吸收
surface abrasion test 表面磨耗试验
surface active agent 表面活性剂
surface activity 表面活性
surface analyzer 表面分析仪
surface cleanliness 表面洁净度
surface combustion 表面燃烧
surface coating 表面涂层
surface condenser 表面冷凝器
surface conditioner 表面调整剂
surface contamination 表面残留物，表面污染
surface corrosion concentration 表面锈蚀[腐蚀]浓度
surface coverage 表面覆盖度
surface denaturation 表面变性
surface density 表面密度
surface diffusion 表面扩散
surface drying 表面干燥
surface echo 表面回波
surface effects 表面效应
surface elasticity 表面弹性
surface energy 表面能
surface enhanced raman spectroscopy (SERS) 表面增强拉曼光谱法
surface excess 过剩浓度，表面余量
surface field 表面（电、磁）场
surface hardening 表面淬火，表面硬化处理
surface heat treatment 表面热处理
surface layer 表面层
surface level 表面能级
surface modification 表面改性
surface morphology 表面形貌
surface noise 表面噪声
surface oxidation 表面氧化
surface phenomenon 界面现象
surface potential 表面电位
surface pressure 表面压力
surface reaction 界面反应
surface relaxation impedance 表面松弛阻抗
surface resistance 表面阻力
surface roughness 表面粗糙度
surface scanner 表面缺陷扫描仪
surface segregation 表面偏析
surface spot 表面瑕疵
surface strain 表面应变
surface strength 表面强度
surface tension 表面张力
surface toughening 表面增韧
surface treatment 表面处理
surface viscosity 表面黏度
surface wave 表面波
surfactant 表面活性剂
surge 波[冲，脉，颤]动
surge chamber 缓冲容器，调压室
surge magnetization 脉冲磁化
surplus 剩余的，过剩的
surplus sensitivity 灵敏度余量
susceptiveness 灵敏度，磁化率
suspend matter 悬浮物
suspended 悬浮物
suspended matter 悬浮物，悬浮质，混悬物质
suspending agent 悬浮剂
suspensible 可悬浮[挂]的
suspension 悬浮液
suspension colloid 悬胶质
suspension polymerization 悬浮聚合
suspensoid 悬胶质，悬浮胶体
swab 擦，抹
swabbing 擦，抹，刷涂料
swage 型砧，冲模，旋锻工具；型铁，铁模，用型铁弄弯曲
swaging 挤锻
Sward hardness rocker 斯惠特硬度计

sweep	扫描
sweep range	扫描范围
sweep speed	扫描速度
sweet oil	橄榄油
swell	膨胀
swelling	膨胀［润］；膨胀［肿大，突起］的
swelling agent	膨胀剂
swelling anisotropy	膨胀向异性
swelling fracture	膨胀破裂
swelling pressure	膨胀压力
swelling resistance	抗胀性
swelling volumetric	容积膨胀
swept cathode	移动阴极
swept field	扫场
swept gain	扫描增益
swirly	成涡旋形的
switching potential	换向电位
swivel scan	环绕扫查，旋转扫描
symmetric	对称的，匀称的
symmetric compound	对称化合物
symmetrical	对称的，匀称的
symmetrical molecule	对称分子
symmetrization	均衡，相称，对称性
symmetry	对称，对称性
symmetry axis	对称轴
symmetry factor	对称因子
synchronism	同时性
synchronize	使……同步；同步，同时发生
synchronization	同一时刻，同步
synchroscope	同步示波器
synchrotron	同步加速器
synchrotron radiation	同步辐射
syncrystallization	同结晶作用
syndet	合成洗涤剂
syneresis	脱水收缩
synergism	增效作用
synergism effect	正协同效应
synergist	协同剂，增效剂，配合剂
synergistic	协同的，协作的，协同作用的
synergistic effect	协同效应
synthesis	合成
synthetic	合成物；综合的，合成的，人造的
synthetic detergent	合成洗涤剂
synthetic dye	合成染料
synthetic fiber	合成纤维
synthetic fiber separator	合成纤维隔板
synthetic material	合成材料
synthetic resin paint	合成树脂涂料
synthetic resin separator	合成树脂隔板
system	制度，系
systematic analysis	系统分析
systematic error	系统误差

T

tack dry 指触干燥
tack free 半固化干燥
tack free test 不黏性试验
tack rag varnish 黏灰清漆
tack rag 黏性纱布
Tafel slope 塔菲尔斜率
tailing peak 拖尾峰
take out device 取料装置
take-off valve 输出阀
taking down 下挂
talc 滑石粉
tank gauge 液面计，油箱液位计
tangential displacement 切线位移
tangential temperature 切线温度
tank lining 镀槽内衬
tank voltage 槽电压
tannic acid 丹宁酸，鞣酸
tannin 鞣酸
tanning 鞣制
tanning agents 鞣制剂
tarnish 失去光泽，变暗，褪色
taper 锥形
taper turning 锥度车削
tapered allowance 削尖余量，锥形宽容
tapered terminal post 锥形接线柱
tapping 出钢，开孔，出渣，轻敲声
target 靶，目标
tarnish 失泽
tartaric acid 酒石酸
tautomerism 互变异构现象
tautomerization 互变异构化
tear 开裂，撕裂
tear strength 撕裂强度
tear test 撕裂试验
tearing 裂痕
temper brittleness 回火脆性
technical 技术的
technical requirement 工艺要求

technical service 渗碳箱
technical specification 技术规格，技术说明书
temper brittleness 回火脆性
temper color 回火颜色
temper resistance 耐回火性
temperature coefficient of electromotive force 电动势的温度系数
temperature range 温度范围
tempered martensite 回火马氏体
tempering 回火
tempering crack 回火裂痕
tempering resistance 回火稳定性
tempo printing 移印
tenacity 韧度
tender 投标
tensammetry 表面张力电量法，张力法
tensile creep 拉伸蠕变
tensile elongation 延伸率
tensile impact test 拉伸冲击试验
tensile loading 拉伸负荷
tensile modulus 拉伸模数
tensile properties 抗拉特性
tensile relaxation modulus 拉伸松弛模量
tensile strength 抗拉强度
tensile stress 张应力
tensile test 张力测试，拉伸试验
tension 张力，拉力
tenth-value 十分之一
terminal 端子
terminal block 线弧，接头排 接线盒，接线板，线夹
terminal box 接线盒
tertiary 第三位的
test block 试块，试验台
test coil 检测线圈
test flow chart 测试流程图
test frequency 试验频率

test panel　试验板
test piece　试片，试验样板
test pump　试验泵
test quality level　检测质量水平
test range　探测范围
test ring　试环
test run　试运行
test specimen　试样
test surface　探测面
test tube　试管
testing　测试
tetra-allylammonium　四烷基铵
tetrahedral and square-planar complexes　四面体和平面四边形配合物
texture　组织，构造
tetrahydrofuran（THF）　四氢呋喃，1,4-环氧丁烷
the wave behavior of electrons　电子的波动性
theoretical capacity of active material　活性物质的理论容量
thermal aging　热老化
thermal analyzer　热分析仪
thermal barrier　热屏障，保温层
thermal conductivity detector（TCD）　热导检测器
thermal conductivity　导热性
thermal conductivity factor　导热系数
thermal contraction　热收缩
thermal cycle　热处理工艺周期
thermal cycle test　热循环试验
thermal diffusion　热扩散
thermal endurance　耐热性
thermal expansion coefficient　热膨胀系数
thermal insulation　热绝缘
thermal oxidation　热氧化
thermal oxidation furnace　热氧化炉
thermal physical property tester　热物理性能测定仪
thermal plasticity　热塑性

thermal polymerization　热聚合
thermal refining　调质处理
thermal resolution　热分辨率
thermal shock　热冲击
thermal shock resistance　抗热冲辉
thermal shock test　冷热剧变试验
thermal stress　热应力
thermic wear　热磨损
thermistor　热阻器
thermodynamic acidity　热力学酸度
thermodynamic functions　热力学函数
thermodynamics　热力学
thermo-chemical treatment　化学热处理
thermocouple　热电耦，温差电偶
thermocouple gage　热电偶真空计
thermodynamic control　热力学控制
thermodynamic reversibility　热力学可逆性
thermodynamics of irreversible processes　不可逆过程热力学
thermoechanical treatment　加工热处理
thermogalvanic corrosion　热偶腐蚀
thermogram　热分析图
thermogravimetric analysis　热重量分析
thermogravimetric curve　热重量曲线
thermomechanical treatment（TMT）　形变热处理
thermometer　温度计
thermoosmosis　热渗透
thermo-paint　测温漆
thermoplastic　热塑性的，热塑性塑料
thermoplasticity　热塑性
thermo-process　热加工
thermosetting　热固的，热硬化的
thermosetting paint　热固型涂料
thermosetting plastics　热固性塑料
thermosetting property　热固性
thermosetting（thermoplastic）acrylic paint　热固（塑）性丙烯酸树脂涂料
thermostat　恒温的，自动调温器，温度调节装置

thermowell	温度计保护管
thero expansion	热膨胀
thickeners	增稠剂
thickening	增稠
thickness gauge	测厚仪
thickness measuring device	测厚仪
thickness sensitivity	厚度灵敏度
thin edcoating	泳透性差
thin layer chromatography, TLC	薄层色谱法
thin layer plate	薄层板
thin paint	涂得太薄
thinner	稀料,稀释剂
thio	含硫的,硫代（氧）的
thiocyanate	硫氰酸盐（或酯）
thioester	硫代酸酯
thiol	硫醇
thiol acid	硫羰酸
thiourea	硫脲
thixotropic	触变的
thixotropic penetrant	摇溶渗透剂
thixotropy	（凝胶等所具有的）触变性,摇溶（现象）
thread	螺纹
thread cutting	螺纹切削
threaded pipe	螺纹管
threading	车螺纹
three-dimensional	三维的；立体的
three dimensional painting	立体涂漆
three electrode system	三电极体系
threshold effect	门槛效应,阈效应
through hardening	透淬,穿透淬火,穿透硬化
through heating	穿透加热
through printing	漏印
through transmission technique	透射技术
throughput volume	（离子交换的）总体积
throwing power	分散能力,泳透力
thymol blue	百里酚蓝
tick-mark farside	反面压印
tick-mark nearside	正面压印
tight	密封
time quenching	时间淬火
tin compound	锡化合物
tin foil	锡箔
tin soap	锡皂
tinning	热浸镀锡
tinplate	镀锡板
tinting	调色
tint	染色,着色于；色彩,色泽
tip-enhanced raman spectroscopy	针尖增强拉曼光谱技术
titanum dioxide	二氧化钛
titanium tetrachloride	四氯化钛
titanium trichloride	三氯化钛
titer	滴定量[度],浓度测定
titrametric analysis	滴定分析法
titration	滴定量[度],浓度测定
titration curve	滴定曲线
titration error	滴定误差
titrator	滴定仪
titrimetry	滴定分析
tolerance	误差,公差
toluene	甲苯
tone analysis	色调分析
tool steel	工具钢
top coat	涂面漆
top reciprocator	平面涂装机
topographic contrast	形貌衬度
torque	扭矩
torsion	扭曲
torsion load	扭转载荷
torsion test	扭曲试验
torsion viscometer	扭力黏度计
torsional creep	扭蠕变
total acid	总酸度
total organic nitrogen	总有机氮
total quality management	全面品质管理
total reflection	全反射
total residual chlorine	总有效氯
total voltage	总电压
total weight	总重量

total-reflection X-ray fluorescence spectroscopy 全反射 X 射线荧光光谱
touch up 修补涂装
touch up primer 局部修补底漆
touching tracer 修整追踪装置
toughening mechanism of composite materials 复合材料增韧机制
toughness 韧度，韧性
towing 擦光
toxic 有毒的
trace 痕量
trace amounts 微量，痕迹量
trace concentration 微量浓度
trace determination 微量测定
trace element 痕量元素
trace level 痕量级
transacetalation 缩醛交换
transacetalization 转缩
transamination 转氨基作用
transannular insertion 跨环插入
transannular rearrangement 跨环重排
transducer 转换器，换能器，传感器
transesterification 酯基转移
transfer 输送，转移
transfer coefficient 传递系数
transfer efficiency 涂着效率
transfer feed 连续自动送料装置
transfer molding 转送成形
transformation 变态
transformation involving diffusion 扩散型相变
transformation stress 相变应力
transformation temperature 相变点
transgranular corrosion 穿晶腐蚀
transgranular fracture 穿晶断裂
transgranular stress corrosion cracking 穿晶应力腐蚀断裂
transient 暂态的、短暂的、瞬时的，瞬时现象
transient state 暂态
transition 过渡，转换
transition metal 过渡金属
transition state theory 过渡态理论
translucent media 半透明介质
transmission 透射，发送，传递［输］
transmission coefficient 透射系数，传输系数
transmission densimeter 透射式密度计
transmission efficiency 透射效率
transmission electron microscope 透射型电子显微镜
transmission electron microscopy 透射电子显微术
transmission method 透射方法
transmission point 透射点
transmission technique 透射技术
transmittance 透光率，光透射，透射系数
transmittance spectra 透射光谱
transmitted chains 传动链
transmitted film density 底片透射黑度
transmitted pulse 发射脉冲
transparency 透明度
transparent 透明的
transpassive 过钝化
transport 输送
transport number 迁移数
transverse resolution 横向分辨率
transverse wave 横波
trapping 截留
traveling scan 移动扫描，横向扫描
treated sewage 处理过的污水
trees 树枝状结晶
trench etching 沟渠蚀刻
triangular array 三角形阵列
tribed 三层床
trichloroethylene 三氯乙烯
trichloroethylene clean 三氯乙烯清洗
tricking filter 生物滤池
triene 三烯
trigger (alarm) condition 触发，报警状态
trimethylamine 三甲胺
triethanolamine 三乙醇胺

triethylamine 三乙胺
trimerization 三聚
trim 切边，修边
trimming 整理，修剪，装饰品，配料，去毛边
triphosphate 三磷酸盐
triple layer nickel coating 三层镍
triple traverse technique 三次波法，三次反射技术
triplet 三重峰
trivalent chromium 三价铬
troostite 屈氏体
tropical test 热带气候试验
trouble shooting 故障处理
truck-loading furnace 台车式炉
tuberculation 结瘤腐蚀
tufftride 软氮化，氰化钾盐浴扩散渗氮
tufftride process 软氮化处理

tungsten 钨
tungsten high speed steel 钨高速钢
turbidimeter 浊度计
turbidity 浊度
turn-over 更新期
turning electron numbers over 粒子数反转
tuyere 风嘴，风口
twin boundary 孪晶界
twin crystal 双晶
twinned dendrites 对生枝晶
twinning 双晶作用
twist test 扭转试验
twisting 扭转
two component paint 双组分涂料
two-dimensional 二维的，平面的
two-tone 双色
two-way sort 两档分选
tygon 聚乙烯

U

Ukovic equation 尤考维奇方程式，经典极谱法的扩散电流公式
ulrasonic testing 超声检测
ultimate analysis 元素分析
ultimate biodegradation 最终生物降解
ultimate properties 极限特性
ultimate strain 极限应变
ultimate strength 极限强度
ultimate vacuum 极度真空
ultra filtration (UF) 超滤
ultra filtration equipment 超滤装置
ultra high molecular weight polymer 超高分子量聚合物
ultra low temperature freezer 超低温冰箱
ultra violet curing 紫外线固化干燥
ultracentrifugation 超离心分离
ultracentrifuge 超离心机
ultracentrifuge method 超离心法
ultrafilter 超滤器
ultrafilter membrane 超过滤膜
ultrafiltration 超细过滤，超滤
ultrafine fiber 超细纤维
ultrafine particle 超细粒子
ultrafine powder 超细粉
ultra-high vacuum 超高真空
ultramicro analysis 超微分析
ultramicro crystal 超微晶体
ultramicro chemistry 超微量化学
ultramicron 超微粒子
ultramicroscope 超倍显微镜
ultrared ray 红外线
ultrared ray drying 红外线干燥
ultrasonic cell disruptor 超声破碎仪
ultrasonic cleaning 超声波清洗
ultrasonic equipment 超音波装置
ultrasonic examination 超声波探伤，超声检验
ultrasonic field 超声场
ultrasonic flaw detector 超声探伤仪
ultrasonic flowmeter 超声波流量计
ultrasonic fractography 超声波断层显微分析
ultrasonic frequencies 超音频率
ultrasonic inspection 超声波探伤法
ultrasonic leak detector 超声波检漏仪
ultrasonic machining 超音波加工
ultrasonic microscope 超声显微镜
ultrasonic noise level 超声噪声电平
ultrasonic spectroscopy 超声频谱
ultrasonic test 超声检测
ultrasonic testing system 超声检测系统
ultrasonic thickness gauge 超声测厚仪
ultrasonic wave 超声波
ultraviolet absorber fixative 紫外线吸收固定剂
ultraviolet absorption 紫外光吸收
ultraviolet absorption spectra 紫外光吸收光谱
ultraviolet and visible spectrophotometry 紫外-可见分光光度法
ultraviolet detector 紫外检测器
ultraviolet irradiation 紫外光照射
ultraviolet photography 紫外线照相
ultraviolet photoelectron spectroscopy 紫外光电子能谱
ultraviolet radiation 紫外辐射
ultraviolet ray 紫外线
ultraviolet ray microscope 紫外线显微镜
ultraviolet spectrophotometry 紫外线分光光度测定法
ultraviolet stabilizer 紫外线稳定剂
ultraviolet 紫外线
ultrviolet absorber 紫外线吸收剂
umpire analysis 仲裁分析
unbranched molecule 无支链分子
undecanal 十一醛

undecane	十一烷
undecanoic acid	十一酸
undecyl alcohol	十一醇
undecylenic acid	十一碳烯酸
undecylenic alcohol	十一碳烯醇
undecylic acid	十一酸
under annealing	不完全退火
under baking	未烘干透
under cooling	过冷
under development	显影不足
under etching	蚀刻不足
under spray	喷油不足，起牙边
undercoat	底涂
underexposure	照射不足
underfilm corrosion	膜下腐蚀
underglaze color	釉底颜料
underground water	地下水
underpotential deposition	欠电势沉积
under-voltage	欠压
uneven dyeing	染色不匀
uneven spray	喷油表面不均
unevenness	凹凸，不均匀
unfitness of butt joint	错边量
uniaxial crystal	单轴晶体
uniaxial orientation	单轴取向
unidirectional	单向的，单向性的
uniform corrosion	均匀腐蚀
uniform thickness distribution	厚度分布均匀
uniformity	均质性
uniformity coefficient	均匀系数
uniformity of film	膜均一性
unimolecular layer	单分子层
unimolecular reaction	单分子反应
uninflammability	不燃性
union colorimeter	联合比色计
unionized free ammonia	游离氨
unit activity	单位活性
unit cell	晶胞
unit matrix	单位矩阵
unit operation	单元操作
unit process	单元过程
unit time	单位时间
univalent	单价的
univariant system	单变物系
universal bridge	万用电桥
universal constant	通用常数
universal indicator	通用指示剂
unpaired electron	不成对电子，孤电子
unqualified products	不合格品
unsaponifiable matter	非皂化物
unsaturated bond	不饱和键
unsaturated compound	不饱和化合物
unsaturated hydrocarbon	不饱和烃
unsaturated polyester paint	不饱和聚酯涂料
unsaturated polyester	不饱和聚酯
unsaturated solution	不饱和溶液
unsaturation	不饱和
unshared electron pair	未共享电子对
unsharpness	不清晰，不清晰度
unstable compound	不稳定化合物
unstable equilibrium	不稳定平衡
unsteady state	非稳定态
unsymmetrical	不对称的
unusual valency	异常原子价
unvulcanized rubber	未硫化橡胶，生胶
upsetting	锻粗加工
upsiding down edges	翻边
upward	向上
uranium	铀
uranium dioxide	二氧化铀
urea adduct	尿素加合物
urea	尿素
urea anhydride	尿素酐，氨基氰
urea chloride	氨基甲酰氯
ureide	酰脲
urethane resin	聚氨酯树脂
urface roughening	橘皮状表皮皱褶
urotropine	乌洛托品，六亚甲基四胺
ursolic acid	乌索酸
urushiol	漆酚

urushiol resin paint 漆酚树脂漆
used oils 废油
useful density range 有效密度范围

UV absorbers and light stabilizers 紫外光吸收剂和光稳定剂
uviol glass 透紫外线玻璃

V

vacancy 空位
vacancy band 空带
vacant electron site 电子空位
vacuum 真空
vacuum apparatus 真空装置
vacuum carbonitriding 真空渗碳氮化
vacuum carburizing 真空渗碳法
vacuum cassette 真空暗盒，真空盒
vacuum condensing point 真空冷凝点
vacuum crystallizer 真空结晶器
vacuum deposition 真空镀
vacuum desiccator 真空干燥器
vacuum distillation 真空蒸馏
vacuum dryer 真空干燥器
vacuum drying 真空干燥
vacuum evaporation coating 真空镀膜
vacuum evaporation 真空蒸发
vacuum evaporator 真空蒸发器
vacuum filter 真空过滤器
vacuum filtration 真空过滤
vacuum furnace 真空炉
vacuum hardening 真空淬火
vacuum heat treatment 真空热处理
vacuum nitriding 真空氮化
vacuum oven 真空烘箱
vacuum plating 真空镀膜
vacuum pump 真空泵
vacuum technique 真空技术
vacuum test 真空试验
valence 化合价
valence band 价带
valence bond 价键
valence bond theory 价键理论
valence electron 价电子
valence state 价态
valeraldehyde 戊醛
valeric acid 戊酸

value of lightness 明亮度
valve 阀，闸门
Van der Waals force 范德瓦耳斯力
Van der Waals molecule 范德瓦耳斯分子
vanadate 钒酸盐
vanadic acid 钒酸
vanadium 钒
vanadium compound 钒化合物
vanillin 香草醛，3-甲氧基-4-羟基苯甲醛
vapor bath 蒸汽浴
vapor degreasing 蒸气脱脂
vapor deposition 气相沉积
vapor diffusion 蒸汽扩散
vapor drying equipment 蒸汽干燥装置
vapor liquid equilibrium 气液平衡
vapor lock 汽封，汽塞现象
vapor loss 蒸汽损失
vapor nozzle 蒸汽喷嘴
vapor phase 汽相
vapor pressure 蒸汽压
vapor pressure thermometer 蒸汽压式温度计
vaporization 蒸发，汽化
vaporization heat 蒸发热
vaporizer 汽化器，蒸馏器
vapor degreasing 蒸汽除油，蒸汽脱脂
variable angle probe 可变角探头
variable resistor 可变电阻
varnish 清漆，亮光漆
vector 矢量，向量
vegetable acid 植物酸
vegetable gum 植物胶
vegetable oil 植物油
vehicle 载体
velocity 速度
velocity constant 速度常数
vent 放气孔，通风孔

vent hole 通气孔
venter 排气风扇
ventilation 通风设备
ventilator 换气扇
verdigris 铜绿，碱式碳酸铜
verification 检验
verify 校验
vertical 垂直的，立式的
vertical linearity 垂直线性
vertical location 垂直定位
vertical panel 竖直面板
verticality 铅垂度，垂直性
vessel 容器
vibration test 振动试验
vibrating ball mill 振动球磨机
vibrating conveyer 振运机
vibrating screen 振动筛
vibrating sieve 振动筛
vibrating electrode 振动电极
vibration 振动，震荡
vibration damping 振动衰减
vibration frequencies 振动频率
vibration insulators 隔振器
vibration stopper 减震器
vibrational relaxation 振动弛豫
vickers hardness 维氏硬度
vickers hardness test 维氏硬度试验
viewing area 观察区域（评片灯上的评片观察区域或窗口）
vinegar 醋
vinyl acetate 醋酸乙烯酯
vinyl alcohol 乙烯醇
vinyl chloride copolymer 氯乙烯共聚物
vinyl chloride 氯乙烯
vinyl cyanide 丙烯腈
vinyl ester 乙烯基酯
vinyl ether 乙烯醚
vinyl isobutyl ether 乙烯基异丁基醚
vinyl polymer 乙烯基聚合物
vinyltoluene 甲苯乙烯
virgin paint 原漆

virtual image 虚拟图像
viscometer 黏度计
viscosity 黏度
viscosity index 黏度指数
viscosity measurements 黏度测量
viscosity modifiers 黏度改变剂
viscous flow state 黏流态
viscous flow temperature 黏流温度，软化温度
viscous liquid 黏性液体
vises 虎钳
visible light 可见光
visible rays 可见光线
visual colorimeter 目测比色计
visual inspection 肉眼检验，外观检验
visual method 目视法
vitamin pp 烟酰胺
vitamin 维生素
vitrification 玻璃化，透明化
void 空隙，空洞
voidage 空隙度
volatibility 挥发性
volatile 挥发性的
volatile acid 挥发酸
volatile matter 挥发物
volatile organic contaminant 挥发性有机污染物
volatile phenol 挥发性酚
volatile solvent 挥发性溶剂
volatility 挥发性，挥发度
volatilization 挥发
volatilization method 挥发法
volta cell 电池
voltage 电压
voltage drop 电压降
voltage stabilizer 电压稳定器
voltage threshold 阈值电压
Voltaic Cell 伏特电池
voltammetric analysis 伏安分析法
voltammetric titration 伏安滴定法
voltammetry 伏安法

voltmeter 电压表，电压计
volume 容积
volume current density 体积电流密度
volume density 体积密度
volume elasticity 体积弹性
volume flow 容积流量
volume fraction 容积比
volume percent 容积百分数
volume resistance 体积阻力
volume viscosity 体积黏性
volumeter 体积计
volumetric analysis 容量分析法，滴定分析法
volumetric determination 容量测定
volumetric factor 容量因数
volumetric flask 容量瓶
volumetric method 容量法
volumetric solution 滴定液
vortex 涡流
vulcanite 硬橡胶
vulcanizate 硫化橡胶
vulcanization 硫化

W

wafer　圆片，晶片
Wall-jet electrode　壁面-射流电极，壁喷电极
Wall-tube electrode　壁面-管道电极
warm color　暖色
warm forging　温锻
warpage test　翘曲试验
warpage　翘曲
washability　可洗性
water-washable penetrant　可水洗型渗透剂
washing　洗涤
washing agent　去垢剂，洗涤剂
washing bottle　洗涤瓶
washing effect　洗涤效应
washing liquid　洗涤液
washing powder　洗涤（剂）粉
washing soap　洗衣皂
washing soda　洗涤碱
washing solvent　清洗用溶剂
waste　废物
waste acid　废酸
waste electrical and electronic equipment　废旧电器，电子废弃物
waste fluid separator　废液自动分离机
waste gas　废气
waste lye　废碱液
waste oil　废油
waste oil regeneration　废油再生
waste product　废产物，废品
waste water　废水
watch glass　表面皿
water absorption　吸水率
water bath　水浴
water bath kettle　水浴锅
water break test　水膜破坏试验，破水试验（测定液体的润湿性）
water break　不连续水膜
water bubbling　水泡

water chiller　水冷器
water circulation　水循环
water color　水合颜料
water column coupling method　水柱耦合法
water conditioning　水质调节
water content　含水量
water cooler　水冷却器
water cooling　水冷却
water curtain　水幕
water dispersible paint　水分散型涂料
water gage　水位指示器
water glass　水玻璃，硅酸钠
water-line paint　水线涂料
Walden inversion　瓦尔登反转
water paint　水性漆
water permeability　渗水性
water proof abrasion sandpaper　耐水性砂纸
water purification　水的净化
water purifier　净水器
water purifying plant　净水设备
water quenching　水淬火
water resistance　耐水性
water seal　水封
water softener　软水剂
water softening　水的软化，球化水冷退火
water soluble functional coatings　水溶性涂膜
water soluble oil　水溶性油
water soluble polyesters　水溶性聚酯
water soluble polymers　水溶性聚合物
water soluble resin paint　水溶性树脂涂料
water soluble resin coating　水溶性树脂涂层
water solution　水溶液
water spot　水斑，水渍
water tank　水槽
water tolerance　水容限，耐水性，吸水

量，允许含水量
water toughening 水韧处理
water transfer printing 水转印
water treatment 水处理
water vapor 水蒸气
water vapor permeability 水汽透性
water washing spray booth 水洗式喷漆室
waterline corrosion 水线腐蚀
waterproof 耐水性
waterproof agent 防水剂
waterproof paint 防水涂料
water-proofing agent 防水剂
waterproofing 防水性
water soluble dye 水溶性染料
wave 波，波纹
wave crest 波峰
wave equation 波动方程
wave form 波形
wave front 波前，波阵面
wave function 波动函数
wave guide 离子波束引导管，导波管
wave guide acoustic emission 声发射波导
wave length 波长
wave number 波值
wave node 波节
wave scanner 波形扫描器
wave train 波列
wave trough 波谷
wave length dispersion 色散
wavelet transform 离散小波变换
waviness 波痕，波形起伏
wax 蜡
wax cement 蜡胶黏剂
weak acid 弱酸
weak band 弱带
weak base 弱碱
weak complex compound 弱络合物
weak electrolyte 弱电解质
wear 磨耗
wear failure 磨损失效
wear process 磨损过程

wear rate 磨损速度，磨损率
wear resistance 耐磨性
wearability 耐磨损性
weather 气候
weather resistance 耐候性，耐风化
weatherability 耐候性
weathering 风蚀
weathering bloom 风化作用
weathering test 风化试验，耐候性实验
weatherometer 人工老化试验机
weatherproof 耐候性
webbing 熔塌
wedge 斜楔
weighing 称量
weighing accuracy 称量准确度
weight 重量
weighting 称重
weighing form 称量形式
weight determination 重量测定
weight distribution 重量分布
weight loss on heating 加热失重
weight energy density 质量能量密度
weight percent 质量百分数
weld decay 焊接腐蚀
weld mark 焊痕
welding 焊接
welding flux 焊剂
welding inspection 焊接检验
welding line 焊缝
welding procedure 焊接工艺
welding rod 焊条
weldment 焊接件
well 井
Weston normal cell 韦斯顿标准电池
wetting 润湿
wettability 可润湿性
wetting agent 润湿剂
wetting action 润湿作用
wetting power 润湿力
wetting tendency 润湿倾向
wet cleaning equipment 湿式洗涤装置

wet coating	湿涂膜
wet contamination	湿污染
wet etching system	湿式蚀刻系统
wet extrusion	湿法挤压
wet film	湿涂膜
wet film thickness gauge	湿漆膜厚度量计
wet fixation	湿法固定
wet granulation	湿法制粒
wet grinder	湿磨机
wet grinding	湿磨
wet grinding mill	湿磨机
wet milling process	湿磨法
wet metallurgy	湿法冶金
wet on wet	湿碰湿工艺
wet out rate	浸润速率
wet rotary mill	湿式转磨碎机
wet sanding with gasoline	汽油打磨
wet slurry technique	湿浆（膏剂）法
wet vapor	湿蒸汽
wheel head	磨轮头
wheel spindle	磨轮轴
wheelabrator	砂粒喷磨机
whetstone	磨石
whirler	旋转涂膜机
whiskers	晶须
white noise level	白噪声水平
white pigment	白色颜料
white titanium pigment	钛白
white wash	白浆
white wax	白蜡
white zinc-plating	蓝白锌电镀
whitening	白化
whole pipet	移液吸管
wide plastics	通用塑料
width tolerances	幅宽公差
wind	卷绕
winding accuracy	绕组精度
wire brush	钢丝刷
wire cutters	剪线钳
wire electrode beam	丝束电极
wire gauze	铁丝网
wire	金属丝，电线
wire sieve	金属丝网筛
wiring	抽线加工
withstand voltage test	耐（电）压试验
witnessed inspections	现场检测
wolfram	钨
wolframate	钨酸盐
wolframic acid	钨酸
wood grain	木纹
wood spatula	木刮勺
woody fracture	木纹状断口
work brittleness	加工脆性
work clamp	工件夹
work damage layer defect	加工层损伤缺陷
work hardening	加工硬化
work function	功函数
work of deformation	变形功
work of friction	摩擦功
workability annealing	改善加工性的退火
workability	可加工性，可变形性
working electrode	工作电极
working load	工作荷载
working operation	工序
workpiece	工件
works process specification	典型工艺规程，工艺说明书
workshop test	现场试验
worst dynamic load	危险动负载
woven cloth tube	纺布管
woven separator	织物隔板
wrapping test	缠绕试验，包装试验
wrench	扳手
wrinkle	皱纹，起皱
wrinkle recovery	旧痕复原，折痕回复角试验
wrinkle resistance	耐皱度，抗皱性
wrought	制造的，加工的，经装饰的，（金属）锤打成形的
wrought aluminium alloy	锻造铝合金

wrought iron 熟铁，锻铁
wrought steel 熟钢
wrought zinc alloy 变形锌合金

wurtzite structure 纤维锌矿型组织
wustite 维氏体

X

X radiation 伦琴射线，X 射线辐射
X ray diffraction (XRD) X 射线衍射
X ray diffraction analysis X 射线衍射分析
X ray diffration pattern X 射线衍射图
X ray fluorescence analysis X 射线荧光分析
X ray goniometer X 射线测角仪
X ray intensity X 射线辐射强度
X ray interferometer X 射线干涉仪
X ray microanalyser X 射线微区分析器
X ray photoelectron spectroscopy X 射线光电子能谱学
X ray protective glass 防 X 射线玻璃
X ray spectrograph X 射线摄谱仪
X ray spectrometer X 射线分光计
X ray spectrometry X 射线光谱分析法
X ray spectrophotometer X 射线分光光度计
X ray spectroscopic analysis X 射线光谱分析
X ray spectroscopy X 射线分光术
X ray spectrum X 射线光谱
X ray tube X 射线管
X ray crack detector X 线裂痕检查器

X ray diffractometer X 射线衍射仪
X ray flaw detector X 射线探伤机
X ray fluorescence spectroscopy X 射线荧光光谱学
X ray fluorometry X 射线荧光分析法
X ray inspection X 射线检查
X ray radiography X 射线照相
X ray shield X 射线保护屏
xeno- 【构词成分】外，异物
xenon 氙
xenon light source 氙光源
xenon lamp 氙气灯
xerogel 干凝胶
xerography 静电印刷术，干印术
xeroradiography 静电影像射线照相术，干法射线照相术
xerox 硒鼓复印机
xylan 木聚糖
xylene 二甲苯
xylenol 二甲苯酚
xylenol orange 二甲酚橙
xylol 二甲苯
xylonite 赛璐珞，硝酸纤维素塑料

Y

yard 码（英长度单位，1码＝0.9144米）
yardstick 码尺
yaw 偏转
yellow pewter 低锌青铜
yield 屈服点，极限，弯曲，凹陷
yield ratio 屈强比
yield band 屈服带
yield limit 屈服极限
yield point 屈服点
yield region 屈服区
yield strain 屈服应变
yield strength 屈服强度，应力强度
yield stress 屈服应力，屈服点
yield temperature 屈服温度，流动温度
yielding 易弯曲的，屈服性的，流动性的，可（易）变形的，塑性变形的，击穿的
ynamine 炔胺
Young's modulus 杨氏模数，弹性模数
Young's equation 杨氏方程
ytterbia 氧化镱
ytterbic 含镱的
ytterbium 镱
yttria 氧化钇
yttric 含钇的，钇的
yttriferous 含钇的
yttrious 钇的
yttrium 钇
yttrium chloride 氯化钇
yttrium nitrate 硝酸钇
yttrium oxide 氧化钇
yttrium sulfate 硫酸钇

Z

Zamak 锌基压铸［锻］合金
zapon 硝化纤维清漆，硝基清漆
zeitter-ion 两性离子
zero activity 零活度
zero adjustment 零位调整
zero defect 零缺点
zero drift 零点漂移
zero level 零水准
zero resistance amperometer 零电阻电流计
zero-working 0℃以下的塑性变形
zeolite 沸石
zeolite sorption pump 分子筛吸附泵
zeyssatite 硅藻土
zig quenching 夹具淬火
zigzag 交错
zigzag dislocation Z字形位错
zinc air batteries 锌空电池
zinc base alloy 锌基合金
zinc bath 镀锌槽
zinc blende structure 闪锌矿型组织
zinc chloride batteries 银氯化物电池
zinc covering 包锌
zinc dross 锌渣
zinc dust 锌粉
zinc impregnation 渗锌，锌化
zinc melting unit 熔锌设备
zinc silicate 硅酸锌
zinc removal treatment 除锌处理
zinc slab 锌锭
zinc stearate 硬脂酸锌
zinc-bearing 含锌的
zinc-coated wire 镀锌钢丝
zinc-rich paint 富锌涂料
zinc-manganese phosphate coating 锌锰系磷化膜
zinc-metal sheet 锌片
zincate 锌酸盐
zincative 负电的
zincic （含）锌的
zinciferous 含［生，产］锌的
zincification 包锌，加锌
zincify 镀锌，包锌，在……上镀以锌
zincilate 锌淬，含锌粉
zincity 镀锌
zincked sheet 镀锌钢板
zincky （含，似）锌的
zincode （电池的）锌极
zincoid 锌的，似锌的
zincolith 白色颜料，锌白
zincous 阳极的，含锌的，锌的
zincsludge 锌泥渣
zinkify 包［镀］锌
zircite 氧化锆
zircon 锆石
zirconate 锆酸盐，锆酸根
zirconia 氧化锆
zirconic 锆的，含锆的
zirconite 褐锆石，锆英石
zirconium 锆
zirconium alloy 锆合金
zirconium steel 锆钢
zirconyl 氧锆基
zonal 带状的，形成地带的
zonal curve 晶带曲线
zonal growth （结晶的）晶带成长
zonal structure 带状组织
zonal texture 带状织构
zone 区，段，带，层，（结晶的）晶带
zone axis 晶带轴
zone leveling 逐区致匀，区域致匀
zone refining 区域精炼
zone structure 带状组织，带状结构
zone travel 熔区移动
zone refining 区域精炼
zone-segregation 熔区偏析
zone-void 熔区空段

zoop 调制噪声
zwitterion 两性离子
zwitterionic 两性离子的

zyglo 荧光透视法，荧光探伤器
zymo- 【构词成分】酶，发酵
zymotic 发酵的

附　　录

附录1　化学元素名称英汉对照表（原子序数100号之前）

原子序数	元素符号	英文名称	中文名称	原子序数	元素符号	英文名称	中文名称
1	H	hydrogen	氢	30	Zn	zinc	锌
2	He	helium	氦	31	Ga	gallium	镓
3	Li	lithium	锂	32	Ge	germanium	锗
4	Be	beryllium(glucinium)	铍	33	As	arsenic	砷
5	B	boron	硼	34	Se	selenium	硒
6	C	carbon	碳	35	Br	bromine	溴
7	N	nitrogen	氮	36	Kr	krypton	氪
8	O	oxygen	氧	37	Rb	rubidium	铷
9	F	fluorine	氟	38	Sr	strontium	锶
10	Ne	neon	氖	39	Y	yttrium	钇
11	Na	sodium	钠	40	Zr	zirconium	锆
12	Mg	magnesium	镁	41	Nb	niobium(columbium)	铌
13	Al	aluminium	铝	42	Mo	molybdenum	钼
14	Si	silicon	硅	43	Tc(Ma)	technetium(masurium)	锝
15	P	phosphorus	磷	44	Ru	ruthenium	钌
16	S	sulfur	硫	45	Rh	rhodium	铑
17	Cl	chlorine	氯	46	Pd	palladium	钯
18	Ar	argon	氩	47	Ag	silver	银
19	K	potassium	钾	48	Cd	cadmium	镉
20	Ca	calcium	钙	49	In	indium	铟
21	Sc	scandium	钪	50	Sn	tin	锡
22	Ti	titanium	钛	51	Sb	antimony	锑
23	V	vanadium	钒	52	Te	tellurium	碲
24	Cr	chromium	铬	53	I	iodium(iodine)	碘
25	Mn	manganum(manganese)	锰	54	Xe	xenonum(xenon)	氙
26	Fe	iron	铁	55	Cs	Caesium(cesium)	铯
27	Co	cobalt	钴	56	Ba	barium	钡
28	Ni	nickel	镍	57	La	lanthanum	镧
29	Cu	copper	铜	58	Ce	cerium	铈

续表

原子序数	元素符号	英文名称	中文名称	原子序数	元素符号	英文名称	中文名称
59	Pr	praseodymium	镨	80	Hg	hydrargyrum(mercury)	汞
60	Nd	neodymium	钕	81	Tl	thallium	铊
61	Pm(Il)	promethium(illinium)	钷	82	Pb	plumbum(lead)	铅
62	Sm	samarium	钐	83	Bi	bismuth	铋
63	Eu	europium	铕	84	Po	polonium	钋
64	Gd	gadolinium	钆	85	At	astatium	砹
65	Tb	terbium	铽	86	Rn(Nt)	radon(niton)	氡
66	Dy	dysprosium	镝	87	Fr	Francium	钫
67	Ho	holmium	钬	88	Ra	radium	镭
68	Er	erbium	铒	89	Ac	actinium	锕
69	Tm	thulium	铥	90	Th	thorium	钍
70	Yb	ytterbium	镱	91	Pa	protactinium(protoactinium)	镤
71	Lu(Cp)	lutecium(cassiopeium)	镥	92	U	uranium	铀
72	Hf(Ct)	hafnium(celtium)	铪	93	Np	neptunium	镎
73	Ta	tantalum	钽	94	Pu	plutonium	钚
74	W	wolfram(tungsten)	钨	95	Am	americium	镅
75	Re	rhenium	铼	96	Cm	curium	锔
76	Os	osmium	锇	97	Bk	berkelium	锫
77	Ir	iridium	铱	98	Cf	californium	锎
78	Pt	platinum	铂	99	Es(An)	einsteinium(athenium)	锿
79	Au	aurum(gold)	金	100	Fm(Ct)	fermium(centurium)	镄

附录2 常见缩写词

缩写词	英文全称	中文释义
A. C.	alternating current	交流电
AAS	atomic absorption spectrometry	原子吸收分光光度法,原子吸收光谱
ABS	acrylonitrile butadiene styrene resin	丙烯腈-丁二烯-苯乙烯树脂
AE	acoustic emission	声发射
AEC	alcohol ether carboxylate	醇醚羧酸盐
AES	Auger electron spectroscopy	俄歇电子能谱
AFM	atomic force microscope	原子力显微镜
AISI	American iron and steel institute	美国钢铁学会
AP	alkyl phosphate	烷基磷酸酯
APG	alkyl polyglycoside	烷基糖苷
ASD	anodic spark deposition	阳极火花沉积
ASTM	American society for testing and materials	美国材料与试验协会
ASV	anodic stripping voltammetry	阳极溶出伏安法
BCC	body-centered cubic	体心立方
BCT	body-centered tetragonal	体心正方
BOD	biochemical oxygen demand	生化需氧量
CA	cellulose acetate	醋酸纤维素
CASS	copper accelerated acetic acid salt spray test	铜盐加速乙酸盐雾实验
CCE	carbon adsorption chloroform extraction	碳吸附-氯仿萃取物
CCT	compound corrosion test	复合腐蚀试验
CCT	continuous cooling transformation	连续冷却转变
CDP	compositional depth profile	成分深度剖析
CE	capillary electrophoresis	毛细管电泳
CE	counter electrode	辅助电极
CGE	capillary gel electrophoresis	毛细管凝胶电泳
CGL	continuous galvanizing line	连续热镀锌线
CMA	cylindrical mirror analyser	筒镜型能量分析器
CMC	critical micelle concentration	临界胶束浓度
CMC	sodium carboxymethyl cellulose	羧甲基纤维素钠
CME	chemically modified electrode	化学修饰电极
CNC	computerized numerical control	电脑数值控制
COD	chemical oxygen demand	化学耗氧量
CPR	corrosion penetration rate	腐蚀速率
CV	cyclic voltammetry	循环伏安法
CVD	chemical vapor deposition	化学蒸镀,化学气相沉积
CZE	capillary zone electrophoresis	毛细管区带电泳法
D. C.	direct current	直流电
DCP	direct current plasma emission spectrometer	直流等离子体发射光谱仪

DME	drooping mercury electrode	滴汞电极
EC	electrolytic colouring	电解着色
EDM	electrical discharge machining	放电加工
EDS	energy disperse spectroscopy	能谱仪
EDTA	ethylenediamine tetraacetic acid	乙二胺四乙酸
EELS	electron energy loss spectroscopy	电子能量损失谱
EIA	energetic-ion analysis	荷能离子分析
ELM	emulsion liquid membrane	乳化液膜
EMF	electromotive force	电池电动势
EMIRS	electrochemically modulated infrared spectroscopy	电势调制红外反射光谱学
EN	electrochemical noise	电化学噪声
ENA	electrochemical noise analysis	电化学噪声分析
EPMA	electron probe microanalysis	电子探针显微分析,电子探针微量分析
EPR	electron paramagnetic resonance	电子顺磁共振
ESCA	electron spectroscopy for chemical analysis	化学分析用电子能谱
ESR	electron spin resonance	电子自旋共振
FABMS	fast atom bombardment mass spectrometry	快原子轰击质谱
FCC	face-centered cubic	面心立方
FD	fluorophotomeric detector	荧光检测器
FFT	fast Fourier transformation	快速傅里叶变换
FRA	frequency response analyzer	频响分析仪
FRP	fiberglass reinforced plastics	玻璃纤维增强树脂
FTIR	Fourier transform infrared spectrometer	傅里叶变换红外光谱仪
FTIR-RAMAN	Fourier transform Raman spectrometer	傅里叶变换拉曼光谱仪
GFC	gel filtration chromatography	凝胶过滤色谱法
FUV	far ultraviolet	远紫外(线,区)
FWDC	full-wave direct current	全波直流
FWHM	full width at half maximum	半高峰宽
GC	gas chromatography	气相色谱法,气相色谱分析
GC-MS	gas chromatography-mass spectrometry	气相色谱-质谱联用
GDMS	glow discharge mass spectrometry	辉光放电质谱
GDOES	glow discharge optical emission spectrometry	辉光放电发射光谱
GDP	galvanostatic double pulse method	恒电流双脉冲方法
GDS	glow discharge spectrometry	辉光放电谱
GPC	gel permeation chromatography	凝胶渗透色谱法
HCP	hexagonal close-packed	密排六方结构
HDPE	high-density polyethylene	高密度聚乙烯
HEISS	high-energy ion-scattering spectrometry	高能离子散射谱
HIPS	high impact polystyrene	高冲击聚苯乙烯,耐冲击聚苯乙烯
HMDE	hanging mercury drop electrode	悬汞电极
HPEC	high performance capillary electrophoresis	高效毛细管电泳法

HPLC	high performance/pressure liquid chromatography	高效液相色谱/高压液相色谱
HPLC-MS	high performance liquid chromatography-mass spectrometry	高效液相色谱—质谱联用
HPMC	hydroxypropyl methylcellulose	羟丙基甲基纤维素
HREELS	high resolution electron energy loss specotroscopy	高分辨电子能量损失光谱
IBA	ion beam analysis	离子束分析
ICP-AES	inductively coupled plasma-atomic emission spectroscopy	电感耦合等离子体原子发射光谱法
IEC	ion exchange chromatography	离子交换色谱法
IEM	ion exchange membrane	离子交换膜
IER	ion exchange resin	离子交换树脂
IGLC	inverse gas liquid chromatography	反相气液色谱
IHP	inner Helmholtz plane	内亥姆霍兹面
IPE	ideal polarized electrode	理想极化电极
IRRAS	infrared reflection adsorption spectroscopy	红外反射吸收光谱法
IRS	internal reflection spectro electrochemistry	内反射光谱电化学
IR-SEC	infrared spectro-electrochemistry	红外光谱电化学
ISC	ion suppression chromatography	离子抑制色谱法
ISE	ion selective electrode	离子选择性电极
K_{sp}	the solubility-product constant	溶度积常数
LCP	liquid crystal polymer	液晶聚合物
LDPE	low-density polyethylene	低密度聚乙烯
LECD	localized electrochemical deposition	局部电化学沉积
LMP	liquid membrane permeation	液膜分离技术
MAR	moving average removal	移动平均值消除法
MDI	minimum detection limit	最小检测限
MECC	micellar electrokinetic capillary chromatography	胶束电动毛细管色谱
MEISS	medium-energy ion-scattering spectrometry	介质能量离子散射谱
Mid-IR	mid-infrared absorption spectrum	中红外吸收光谱
MIP	microwave inductive plasma emission spectrometer	微波等离子体光谱仪
MMC	metal matrix composite	金属基复合材料
MSFTIRS	multi-steps fourier transform infrared spectroscopy	多步骤电势阶跃傅里叶变换红外光谱学
NHE	normal hydrogen electrode	标准氢电极
NMR	nuclear magnetic resonance	核磁共振
NTA	nitrilotriacetic acid	氨三乙酸
OC	oxygen consuming	耗氧量
OHP	outer Helmholtz plane	外亥姆霍兹面
OTE	optically transparent electrode	光透明电极
OTTLE	optically transparent thin-layer electrode	光透明薄层电极
PAA	polyacrylic acid	聚丙烯酸
PAD	pulsed amperometric detection	脉冲电流检测

PAM	polyacrylamide	聚丙烯酰胺
PE	polyethylene	聚乙烯
PEG	polyethylene glycol	聚乙二醇
PEO	polyethylene oxide	聚环氧乙烷
PIC	paired ion chromatography	离子对色谱法
PLA	polylactic	聚乳酸
PMA	polymethyl acrylate	聚丙烯酸甲酯
PMMA	polymethyl methacrylate	聚甲基丙烯酸甲酯,有机玻璃
PP	polypropylene	聚丙烯
PS	polystyrene	聚苯乙烯
PSD	power spectral density	功率谱密度
PTFE	polytetrafluoroethylene(Teflon)	聚四氟乙烯,特氟龙
PVA	polyvinyl alcohol	聚乙烯醇
PVAC	polyvinyl acetate	聚醋酸乙烯酯
PVC	polyvinyl chloride	聚氯乙烯
PVD	physical vapor deposition	物理气相沉积法
PVP	polyvinylpyrrolidone	聚乙烯基吡咯烷酮,聚维酮
PZC	potential of zero charge	零电荷电位
QCM	quartz crystal microbalance	石英晶体微天平
RDE	rotating disk electrode	旋转圆盘电极
RE	reference electrode	参比电极
RoHS	the restriction of the use of certain hazardous substances in electrical and electronic equipment	在电子电气设备中限制使用某些有害物质指令
RVC	reticulated vitreous carbon	网状玻璃碳
SAM	scanning Auger microscope	扫描俄歇显微镜
SCOT	support coated open tubular column	载体涂层毛细管柱
SDP	sputter depth profile	溅射深度剖析
SDS	sodium dodecyl sulfate	十二烷基硫酸钠
SECM	scanning electrochemical microscope	扫描电化学显微镜
SEM	scanning electron microscope	扫描电子显微镜
SERS	surface-enhanced raman spectroscopy	表面增强拉曼光谱技术
SFC	supercritical fluid chromatography	超临界流体色谱法
SIM	selected ion monitoring	选择性离子检测
SIMS	secondary ion mass spectrometry	二次离子质谱
SLM	supported liquid membrane	支撑液膜
SNIFTIRS	substractively normalized interfacial Fourier transform infrared spectroscopy	差减归一化界面傅里叶变换红外光谱学
SNMS	sputtered neutral mass spectrometry	溅射中性粒子质谱
SPM	scanning probe microscopy	扫描探针技术
SSA	spherical sector analyser	球扇型能量分析器
STFT	short-time Fourier transformation	短时傅里叶变换
STM	scanning tunneling microscope	扫描隧道显微镜

STPP	sodium tripolyphosphate	三聚磷酸钠
TCD	thermal conductivity detector	热导检测器
TERS	tip-enhanced raman spectroscopy	针尖增强拉曼光谱技术
THF	tetrahydrofuran	四氢呋喃,1,4-环氧丁烷
TISAB	total ion strength adjustment buffer	总离子强度调节缓冲剂
TMT	thermomechanical treatment	形变热处理
TXRF	total-reflection X-ray fluorescence spectroscopy	全反射X射线荧光光谱
UF	ultra filtration	超滤
UPS	ultraviolet photoelectron spectroscopy	紫外光电子能谱
UV-VIS	ultraviolet and visible spectrophotometry	紫外-可见分光光度法
WE	working electrode	工作电极
WEEE	directive waste electrical and electronic equipment directive	关于报废电子电气设备指令
XANES	X ray absorption near-edge structure	X射线吸收近边界结构
ZRA	zero resistance amperometer	零电阻电流计

附录3 常见的构词形式

形容词词尾	释译	词例
-able(-ible)	表"可能性"	suitable 适当的,相配的;weldable 可焊的;reversible 可逆的
-al	表"……的"	national 国家的
-ant(-ent)		important 重要的,dependent 从属的
-ar(-ary)		circular 圆形的,secondary 次的
-ed		aged 老化的,large-sized 大尺寸的,concentrated 浓的
-en	表"制(质)的"	golden 金色的
-ful	表"充满"	useful 有用的
-ic	表"属于"	atomic 原子的;
-ical	表"性质"	scientifical 合乎科学的,systematical 系统性
-ive	表"性状"	active 活泼的
-less	表"否定"	useless 无用的
-like	表"相似"	glass-like 玻璃似的
-ly		friendly 友好的
-ory	表"性状"	refractory 难熔的
-ous		various 各种的,numerous 许多的
-y	表"性状"	woody 木质的
名词后缀	释译	词例
-ability	表抽象概念,如性质、状态、行为等	workability 可加工性,可变性
-age		voltage 电压
-al		removal 除去
-ance(-ence)		impedance 阻抗,difference 差别
-ancy(-ency)		brilliancy 光辉,出色,efficiency 效率
-cy		accuracy 精确性
-dom		feedom 自由
-hood		likelihood 可能性
-ic(-ics)		acoustic 声学,physics 物理学
-ing		reading 读数
-ion(-tion,sion,xion)		alkalization 碱化,diffusion 扩散,connexion 连接
-ism		syllogism 推论法,演绎 electromagnetism 电磁,电磁学
-ment		movement 运动
-ness		hardness 硬度

名词后缀	释 译	词 例
-ship		relationship 关系
-th		growth 生长
-ty(-ity)		adhesivity 黏附性 acidity 酸度
-ure		mixture 混合物
-y		factory 工厂
-er	表示人或物	worker 工人
-or	表示人或物	activator 活化剂
-ist	表示人	chemist 化学家,药剂师
动词词尾	释 译	词 例
-en	表"使"	broaden 加宽,harden 硬化
-fy		amplify 放大,verify 证实
-ize(-ise)		oxidize 氧化
副词词尾	释 译	词 例
-ly	表方式、程度	automatically 自动地,extremely 极度地
-ward	表方向	backward(s) 向后
-wise	表示方向、样子	clockwise 顺时针方向,likewise 同样地
常见的构词成分及后缀		
-fold	……倍的,成……倍的	three-fold 三倍的,成三倍地
-free	无……的,免……的	oil-free 无油的,rust-free 无锈的
-gram	表记录下的,图谱	spectrogram (光)谱图
-grapy	表记录工具或结果	autograpy 自动记录仪, spectrograph 摄谱仪
-graphy	表根据记录图谱来研究的方法和学术	spectrography 摄谱学
-meter	表计量仪表	spectrometer 分光计,acidimeter 酸比重计
-metry	表计量方法或技术	spectrometry 能谱测定法,acidometry 酸度测定法
-ology	……学(科)	geology 地质学
-proof	防……的	water-proof 防水的,不透水的 acidproof 耐酸的
-scope	表观测仪器	spectroscope 分光镜
-scopy	表观测方法或学术	spectroscopy 光(能、波)谱学
-tight	不透……的	air-tight 不透气的,气密的
-tron	表电子管、仪器、装置	plasmatron 等离子管,等离子流发生器,等离子电焊机
art-	表"技巧、关节,诡计"	artifical 人工的,artless 朴实的,artifice 诡计